THE HABITAT HOME DECORATOR

FLOORS AND FLOORING

OCTOPUS CONRAN

Please note. The photographs have been collected from all over the world to show as varied a range of ideas as possible within the scope of this book, so that not all the items featured are available through Habitat.

First published in 1985 by
Conran Octopus Limited
28-32 Shelton Street
London WC2 9PH

Art editor Jane Willis
Editor Liz Wilhide
Designed by Gunna Finnsdottir
Editorial assistance Gillian Denton
Picture research by Keith Bernstein

ISBN 1 85029 023 7

Typeset by SX Composing Ltd.

Printed and bound in Italy

CONTENTS

DESIGN ELEMENTS

The floor sets the tone in any living space. One-sixth of the surface area of a room, it is literally the base for the interior as a whole, for furniture, ornaments and decoration. No matter how carefully everything else is put together, if the floor is wrong the overall effect will be uneasy and disjointed.

Since the floor plays such a fundamental role in creating the style and atmosphere in a home, there's a good argument for making it your first choice and designing the rest of your scheme around it. Even if you don't have the opportunity to plan a whole room or home from scratch, you can stand back and take a long, critical look at what you already have to see if it can be improved.

Don't just worry about the overall picture – 'God is in the details' as one famous architect observed. Inexpert laying, a poor finish or lack of maintenance will undermine the wisest choice. It's foolish to lay cork and then not bother to seal it properly; to fit a carpet without adequate underlay; to leave the junction between two types of flooring unresolved.

A floor must work harder than any other surface in your home. It must be safe, provide a degree of comfort and withstand a reasonable amount of wear. It may have to act as a sound or heat insulator. When you choose a floor, select a type that is practical and fulfils your particular requirements.

Luckily there is an enormous variety of floors from which to choose – a wide enough range to satisfy both stylistic and practical needs, and, importantly, to accommodate most budgets. Traditional floorings such as carpet are more versatile and hard-wearing than ever before, while utilitarian materials such as vinyl and linoleum are now available in specially designed ranges; so research all the alternatives.

Although the floor surfaces differ in material – quarry tiles, wood and rug – toning colours create a harmonious yet varied whole.

KLEE/KA
LOUISIA

DEFINING SPACE

Left: The raised kitchen area of this dual-purpose room needs a practical, hard-wearing, easy-to-clean floor surface, while a warmer, more welcoming material is more appropriate to dining. The combination of stripped wood and ceramic tiles is ideal, setting out the distinction between the two zones both physically and psychologically.

Right: A floor can bring visual coherence to a room without overwhelming the decor. A natural wood floor links a space that combines living and dining areas; the rich patterned rug defines the eating area. A second rug is placed where the sitting space begins, its rust colour complementing the patterned carpet without vying for visual effect.

By linking or separating different areas in the home, floors have an important role to play in the way space is defined. A single floor-covering throughout an open-plan area will reinforce the seamless, flexible approach to living; in a more traditional interior, with rooms opening off a main hallway, different types of floor in each area will encourage the sense of separation. But both these extremes have their own disadvantages. Open plan can mean clutter and confusion, with nowhere in particular to put things or do things. One type of flooring throughout will only emphasize this ambiguity. In the same way, abrupt changes in flooring from room to room will increase the feeling of boxiness in a suburban home. A good compromise, however your home is arranged, is to plan a coherent scheme that retains the feeling of spaciousness through careful use of coordinated colours and textures, but tempers the uniformity by adding rugs or mats, or by including contrasting borders and margins.

Picking out areas

Many homes contain spaces that must serve several functions, such as living and dining, cooking and dining, working and relaxing. A change in flooring, even a very subtle one, can help to distinguish between these activities or cater for different practical needs.

A kitchen area within a general living space could be tiled with cork or ceramic tiles – more practical surfaces than carpet. A round dining table would look particularly good on a circle of inset tiles.

A study corner can be picked out by standing a desk and chair on a rug. If filing cabinets and other storage are part of

the arrangement, either sisal matting or exposed wooden boards would make a stable base, as well as signalling to other members of the household that the workspace should be treated as a private area.

A hall should be striking in its own right rather than the indeterminate link it so often is. For practical reasons, it is rarely a good idea to extend fitted carpet right up to the front door – a large doormat will keep muddy footprints to a minimum. A classic entrance flooring is black and white tiles; alternatives include washable materials such as rubber or lino in colours that tone with the other floors visible from the hall.

Levels

Changing the level is one way of making a room special. A dining area raised on a platform has dignity and drama – but make sure the platform is large enough! Putting the television on a dais with books and plants would encourage a viewing-room atmosphere. For a quiet reading corner, a raised area with cushions and lights placed low would be very effective. Cheap seating units can be built in wood around the perimeter of the room and then carpeted.

Internal views

Wherever floorings meet, you must think carefully about how the junction will work. Use metal or rubber edging strips to neaten the seams. If a single flooring has been used throughout a house, a fitted carpet, for example, the view from room to room can be bland and monotonous. Mats, rugs and wooden margins on the stairs will give a hint of variety.

COLOUR

Within the context of art and design, the study of colour is immensely detailed and highly technical. Colours vary not only in hue, but also in intensity (pale to deep) and tone (light to dark). Intricate colour wheels and diagrams have been devised to demonstrate these interrelationships. For practical purposes, however, it is more workable to divide colours into warm – yellows, oranges and reds – cool – blues, greens and violets – and neutral – grey, white and black. This approximates more closely to the way we respond to colour in our immediate environment and how we use it in interiors.

Warm shades suggest intimacy and domesticity; cooler ones are perhaps more formal and elegant. Light colours are reflective and open up a space, while darker, richer ones tend to absorb light. Deep, rich shades reflect more colour on to their surroundings than pastel shades.

Since a floor forms such a large part of the surface area of any interior, getting the colour 'right' is vital. But such a decision cannot be taken in isolation – or you may carefully coordinate walls and upholstery only to discover that you can't find a compatible carpet. Always think schematically.

A good starting point is often to examine the contents of your wardrobe. Is there a dominant colour or shade? Which colours do you instinctively respond to? Colour charts can also be very useful. Keep samples of material, pieces of wool, pictures from magazines, any odd item where the colour appeals to you, then try out different combinations. You can also paint different colours on pieces of paper, lay them side by side, stand back, and try to imagine the final effect.

If you decide on a particular shade and cannot find it straight away, don't give up. Carpet manufacturers include most colours in their ranges, but usually in quite different tones and shades so that one range might be soft and misty, another clear and vibrant. Much the same goes for linoleums and vinyls, too. As for floor stains, specialist manufacturers will mix colours for you, or you can buy made-up samples and experiment on scraps of wood before deciding. Don't go shopping without taking a sample with you to check matching and to see how colours will look together – sometimes colours that appear very similar clash unpleasantly.

Above: The pale, neutral tones of the furnishings and walls are repeated in the carpeting. The colours are similar but not identical; the overall effect is subtle but not boring. The more intensely coloured rug creates an island within the larger space – a room within a room.

Left: Warm colours, top row, suggest domesticity, while cool colours, middle row, bring formality, and neutrals, bottom row, offer a restrained and comfortable atmosphere to a room.

Below: Bright, multi-coloured chequered flooring brings this kitchen alive, counteracting white and pale wood units and walls which could have made the room look overly bland.

PATTERN AND TEXTURE

Pattern can be intrinsic – the design woven into a rug for example – or it can arise from the way components are arranged, as in a chequered floor or tiles – or it can be a customized decoration, as in a stencilled border. On the whole, symmetrical patterns are soothing; asymmetrical patterns are livelier and more arresting.

Too much pattern, on the other hand, can make for visual indigestion. Depending on the richness of pattern, patterned floors probably look best with simple wall coverings. However, they may look well with quite ornate or antique furniture. Tubular steel and glass, metal or modern wood furniture, generally looks best with plainer floors. Of course there can be exceptions – some modern furniture may look good on an extravagantly patterned floor – but not usually.

Bricks, tiles and parquet have their own traditional patterns of laying that have evolved over the years because they are pleasing, regular or simply efficient. Tiles can also be laid quite randomly or to point up certain areas. In kitchens, for example, tiles contrasting in colour could be laid where people tend to stand – in front of the cooker, fridge and work surface. Tiles or cut-up sheet linoleum could also be laid in classic arrangements such as borders or more playfully in the form of a child's game such as hopscotch.

Those man-made materials such as vinyl, designed to simulate brick, stone or marble, could also be said to be patterned. The expensive varieties are very successful, although purists may still not be able to bring themselves to use them, despite their considerable advantages.

Texture also has an important role in influencing the atmosphere of an interior. Shaggy carpet or rugs may soften the contours of a bland, regular room and provide a welcome contrast to hard-edged modern furniture. The sheer lustre and richness of wool and the nubbliness of dhurries, kelims, and other handmade items lend interest in the subtlest and most satisfying of ways. Some tiles could be laid upside down for a rougher effect; conversely, choose the smooth type of bricks if you want an urban, rather than rustic, look. For a simple textured surface, the natural charm of sisal, rush and other woven matting cannot be bettered.

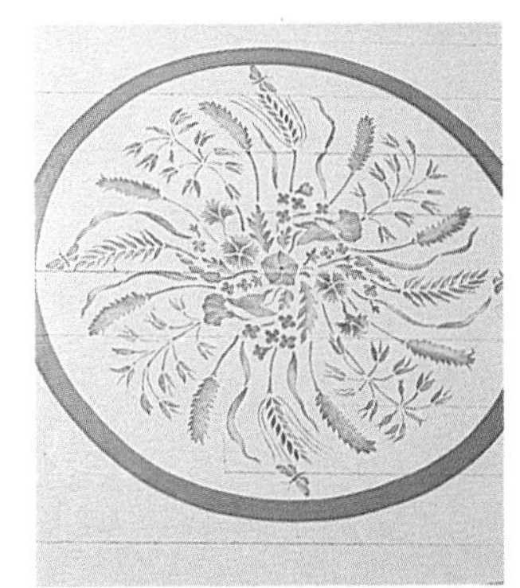

Left: Three different patterns and textures – rag rug, matting and 'carpet' painted on to the hall floor – form a comfortable visual trio, unified by muted colours and polished wood.

Above: Black and white vinyl tiles create pattern without fuss; they are reminiscent of marble.

Right: Painted or stencilled patterns add interest to a plain floor and can carry colour or motif from floor to wall.

STYLE

The contemporary Indian dhurry is quite in keeping with this seventeenth-century farmhouse room; both dhurry and upholstery use re-worked traditional country patterns.

Style tends to be categorized under such popular labels as 'country', 'city' or 'high tech', or with reference to broad historical or cultural periods – Regency, Victorian or Art Deco, for example. Although this may seem rather simplistic, an awareness of how a look can be coordinated in this way can be immensely valuable when designing an interior.

A good starting point can be the period when your home was built. Those who love re-creating historical interiors go to great pains to achieve the original ambience by using contemporary materials and furnishings. In the case of flooring, homes built before 1920 would have featured fired clay, woven materials, timber or stone, in a variety of colours, patterns and textures, according to the date and location. Old pattern books or museum displays can provide inspiration.

Another useful guideline is architectural character. Being sensitive to scale is important. Small townhouses can look overcrowded very quickly, so beware of large patterns. Increase the sense of spaciousness with simple floors – plain carpet, coir, cork, wood – accented with rugs or colourful

Left: As in most classically inspired interiors, the floor is a dominating feature of this bedroom. The cool formality is matched by opulent but restrained furnishings.

Below: Shiny black studded rubber is the perfect base for a high-tech living room, with its formal, but comfortable, grey furniture. The severe effect is softened by the geometric rug.

mats. Avoid materials that seem to demand grander settings.

Modern houses or apartments can look good with a uniform flooring running throughout. A fitted carpet will soften some of the hard edges while at the same time emphasizing the clean lines. Vinyl simulations will be out of place, however – you wouldn't in any case expect to see what is being mimicked.

Don't overdo the country look in a cottage by mixing too many sweetly pretty patterns together. Texture is useful here – rush and coir weaves, rag and braided rugs, woven carpet and wood will all bring out the character of the building.

In the end, however, what counts is that elusive personal touch. Too great a concern for historical accuracy or trying too hard to be appropriate can result in lifeless, though correct, interiors. There's no law that says mansions can't be strewn with coir matting or town drawing rooms tiled in terracotta. As long as your decisions are part of an organized scheme, there's no harm in being adventurous. Expressing your own tastes with flair and originality is true style.

ORIGINAL FLOORS

Marble or mosaic in a hall, quarry tiles or flagstones in a kitchen or lobby, an old wooden floor anywhere, are all surfaces that lend character to an interior. An original floor in such a material should be regarded as an inheritance, especially as it might cost a good deal more new, or even be impossible to reproduce.

The only drawback may be that the floor is in a poor state of repair. Clean it carefully. If you need specialist help – for a damaged tile floor, for example – invest in it; it will probably still cost less than ripping up the floor and laying a new one. 'Period' features are also popular with home buyers and fit in with the style of your home as its builder imagined. Boards with a good surface built up over a long period of time should be left alone – such a patina is priceless. Even if you want to lay carpet, try to leave a border of exposed wood. Similarly, don't cover up old flagstones or quarry tiles or other original hard floors. Even if they are very worn, the uneven aged look is very attractive. In very old houses, bumpy floorboards, boards which slope and stairs which dip are best left unmended and unlevelled unless positively dangerous: again they are part of the charm. As for maintenance, wax polish is better than seal, which might not take on the old surface, or look falsely glossy if it did.

'Original' synthetic floors such as very durable thermoplastic tiles may also be worth retaining. Lift and re-fix loose ones; match and replace damaged ones. If you discover a gem of an elderly linoleum, you may be able to re-use it only if you treat it carefully – it is very prone to cracking. As you lift it, try applying a light oil to the underside.

The strongest visual element of this original 'thirties bathroom is the yellow and white ceramic-tiled floor. The bright, basket-weave pattern would be equally at home in a modern setting, should the fixtures be changed at a later date, and add to the value of the house. They are certainly worth keeping.

PROBLEM AREAS

A narrow corridor is visually widened by using exclusively white tiles on both the walls and floor. A strategically placed mirror heightens the feeling of light and space, and makes what could have been a dark tunnel an exciting and airy place.

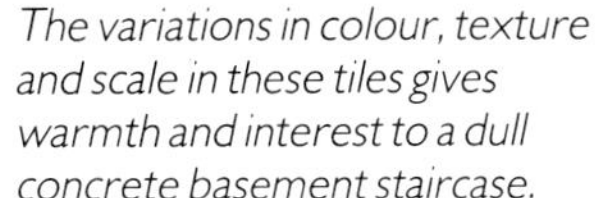

The variations in colour, texture and scale in these tiles gives warmth and interest to a dull concrete basement staircase.

Running carpet over built-in seats in this loft conversion creates a well-lit sitting spot and visually extends the floor space.

Narrow halls A plain, light flooring may broaden the space, but make it overly nondescript. Alternating or chequered tiles smarten such a confined space; borders make it a more interesting tunnel; or an inlaid pattern that incorporates a simple circle in the centre will foreshorten a long narrow hall and break up the monotony.

Stairs A central runner with nicely painted or stained edges is often better than a fitted carpet which may look lumpy at the edges. Runners can be moved up and down to lessen wear, too.

Cellars The usefulness of these areas is much reduced if the environment is so unpleasant you hate going down there. Make sure the floor is clean and damp-free, by treatment and the use of self-smoothing screed. Store things away from floor contact on plastic duckboard or pallets if you are not convinced of the damp-proofing.

Cupboards For smartness and convenience, carpet or lay other flooring material right into cupboards. You could also build a false floor of thin wood and seal or paint it.

Bathrooms Despite all the water, many people want to carpet bathrooms. It is generally not recommended to lay wool carpet which can rot, unless you are very careful and use a thick bathmat. Otherwise, a synthetic fibre carpet is best, specifically one which is resistant to water. However, do not carpet around the toilet. Use an insert of hard flooring material or wooden boards which can be easily cleaned. Do not use glazed ceramic tiles in bathrooms: they are slippery when wet. In a shower install a large good-quality mat.

UNUSUAL SOLUTIONS

Left: Shaped plywood, stained in soft, watery colours, makes a startlingly original floor with a three-dimensional quality. Although time-consuming to achieve, the effect is well worth the effort.

Above: This eye-catching trompe l'oeil *hearth-rug is made of concrete set in the wood flooring. It focuses attention on the fireplace, and unlike a real rug is immune to flying sparks.*

Right: This chequered floor is easily achieved by juxtaposing dark and light cork tiles. They are easy to cut and can make striking borders echoing the main floor pattern.

Below: This highly original floor is made up of rectangles of sheet vinyl in bold colours: a practical, budget-conscious, but stylish solution for a living room needing little other decoration.

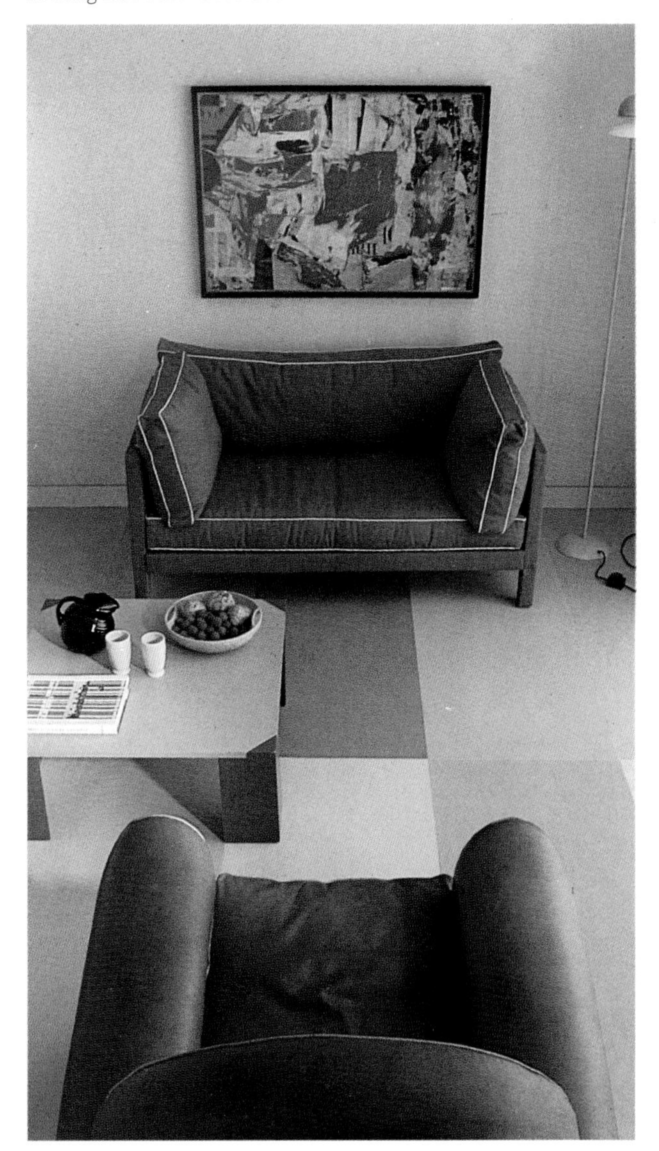

FLOOR-LEVEL LIVING

Left: Oriental-style floor-level living has greatly influenced Western interior design. Here, rice-paper screens, simple floor matting and a low-level Futon sofa bed combine well.

Above: The sea stones are floor level objets trouvés, *but make a useful doorstop too. They draw the eye to the chequered stone floor, which dates from the early eighteenth century.*

In traditional Japanese homes, floors are covered with *tatami* mats (a type of grass mat), wood strips laid in different diagonal sections, slatted wood (particularly in bathrooms) or rounds of wood (*maruta*) embedded in mortar. There are platforms for eating or sleeping and furniture is of the simplest – perhaps just sleeping mats, stowed in a cupboard during the day. Decoration is often floor-level too – small beds of raked gravel or clever stone gardens. This deceptively simple approach, with its spare, serene interiors and emphasis on floor-level living, can be adapted for western use – especially for those who want to create an uncomplicated and tranquil effect without spending a great deal of money.

The floor must have a warm, fairly thick covering and be draught-proof. You can saw legs off inexpensive tables, sofas and chairs to bring them down to floor height; stack books in piles; and invest in chests for clothes storage and cushions and bean bags for sitting.

Many back specialists agree that the ideal sleeping base is firm with a slight degree of give – which perfectly describes a good-quality mattress laid directly on floorboards. Floor-level beds look much less bulky than large bedsteads, and attractive headboards can be improvised from screens, wood strips or padded shapes. Build a platform bed to free more floor space if your living accommodation is tight.

Floor-level ornament may not be practical for homes where there are young children, but such items as big bowls of fruit, pebble arrangements, pot plants in a careful group on a tray or box, a dramatic lamp, statues, driftwood, and pictures and mirrors leaning against a wall can look striking, provided everything else in the room is scaled accordingly. Aim spotlights at these floor-level objects or create pools of light using fittings designed for floor-level use.

Children's rooms also respond well to this type of treatment since children naturally spend a lot of time on the floor. There are carpets and floor coverings printed with board games; cuddly toys look good just piled up in a family group where they can be easily reached; other toys such as a Noah's ark or doll's house set out on the floor would be the right height for small children to play with.

TYPES OF FLOOR

While the colour, texture and style of a floor form the basis of any decorative scheme, at the same time it has far harder practical work to do than any other surface. It must withstand being walked, jumped, danced and sat upon; it must accommodate children learning to crawl and walk, pets and their attendant messes and spills and breakages. It is also the layer that separates you from service pipes, electrical cables and the ground beneath your home. You may agonize over the choice of a dinner service which then sits in a cupboard for much of its life, but a floor is always on duty, and always visible.

Floors aren't cheap either, whether you count the cost in labour or cash, so it is crucial to make a careful and considered choice. A floor is in many ways an investment: you'll want it to last for a reasonable length of time, perhaps through several changes in decor, and you'll certainly expect it to cope with a fair degree of wear. Before making a decision, investigate the full range of possibilities. If you have always assumed that you should have vinyl in the kitchen and carpet everywhere else, think again. That may well be appropriate for your situation – but there are many more possibilities, some of which require only imagination and energy on your part.

A good floor can be the making of a room. Whether it is worn flagstones; waxed, glowing wood; a rich, oriental rug; gleaming marble or deep pile wall-to-wall carpet, it is clear that the experience of flooring, which begins with wiping your shoes on a doormat and ends padding barefoot to bed, is more complex than might first appear and can give a great deal of pleasure if the choices are well made and the floors subsequently well maintained.

The range of floor surfacing materials is enormous; selection should be approached with careful thought and imagination.

ASSESSMENT

Careful assessment both of preferences and requirements is essential before you make any decision about flooring. A new floor can be a hefty investment and will have a dominant effect on any interior; proper consideration of all the alternatives will help to avoid an expensive mistake. A common cause of disappointment is expecting too much from a particular flooring material; to eliminate this problem you should establish exactly what your needs are. The number of people in your household, their ages and pursuits, where the flooring will be laid, the amount you want to spend; all these practical criteria will have a bearing on your choice, together with aesthetic and stylistic issues concerning the type of interior you want to create.

Antique rugs are expensive, but a portable lifetime investment.

Before buying carpeting, or permanent flooring material of any sort, test a good-sized sample in the existing scheme.

Cost

Cost inevitably plays a large part in any decision concerning choice of flooring, but this is not simply a case of reckoning how much you can afford. Decide how long you want the floor to last. Is it more economical to install a cheaper temporary alternative until you can afford the luxury you want?

Some types of floor add to the resale value of your home, but it won't be worth spending a fortune on a new fitted carpet if a move is planned in the near future. A good rug, on the other hand, can be a true investment.

Make sure you have included the full cost of labour, preparation, underlay and fittings in the total cost of the floor. It's all too easy to overlook such additional expenses and they can make substantial inroads into your budget.

Appearance and style

In design terms, your floor should fit in with an overall scheme. There's a lot to be said for making the floor your first consideration and planning the rest of your decoration and furnishing around it. But if this is impractical, you must still take careful account of the character, period and overall style of your home. You don't necessarily have to coordinate flooring with every other element, but if you are going to make a bold statement, it should be a considered – and successsful – one!

Details are important, too. Think about how you are going to treat the junction between the different types of flooring, and between the floor and the walls. Think, too, about how the transitions between the levels or from room to room will work.

Stilettos can damage any wood floor, particularly if veneered.

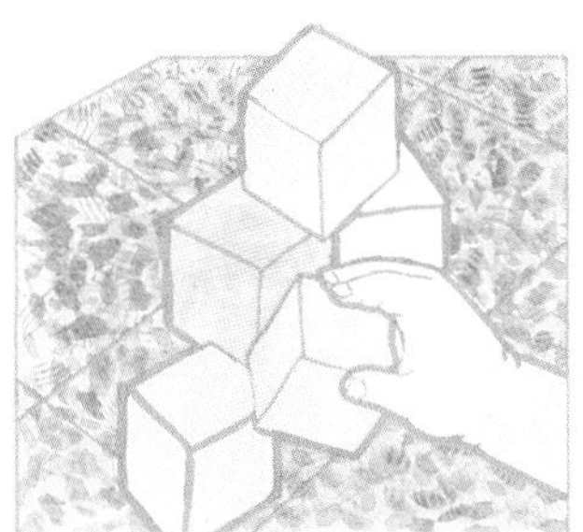

For children, cork is warm, not hard, and impervious to messes.

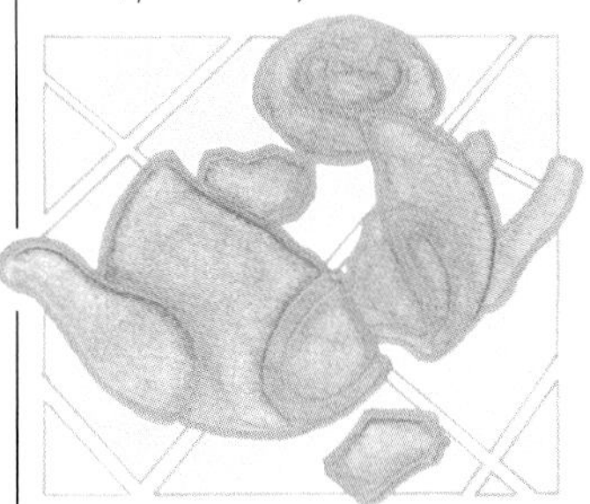

A piece of china dropped on ceramic tiles is certain to break.

If you like going about barefoot, carpet feels warm and soft.

Function

How will the floor be used? Striking the right balance between durability and comfort is important for multi-purpose family rooms; other, more specialized areas in the home may impose certain practical restrictions. Stairs, entrances and halls suffer heavy traffic; kitchens and bathrooms are places where spills and breakages are more likely to occur. Workshops or other areas where garden equipment, motorbikes, bicycles or prams are stored take a lot of punishment.

Users have their own requirements too. Safety, warmth and comfort are particularly important for babies, toddlers and the elderly. Pets can dramatically increase wear and tear. Lifestyle is another factor to bear in mind: do you live *on* the floor? Entertain often? Like to go barefoot?

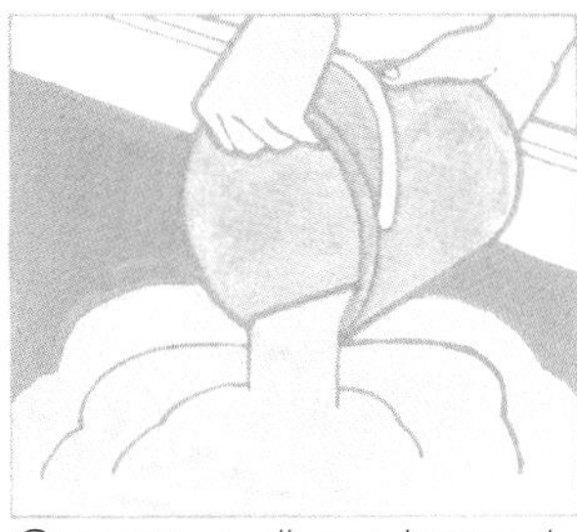

Concrete usually needs screeding before you lay a new floor.

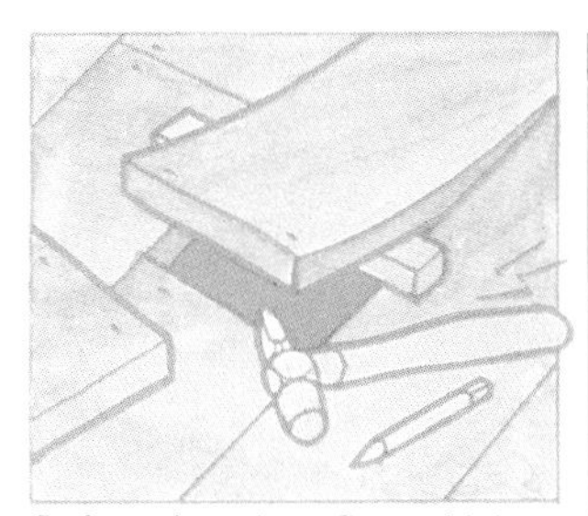

Before choosing a floor, think how much upheaval it will cause.

Sanding causes a lot of dust so start at the top of the house.

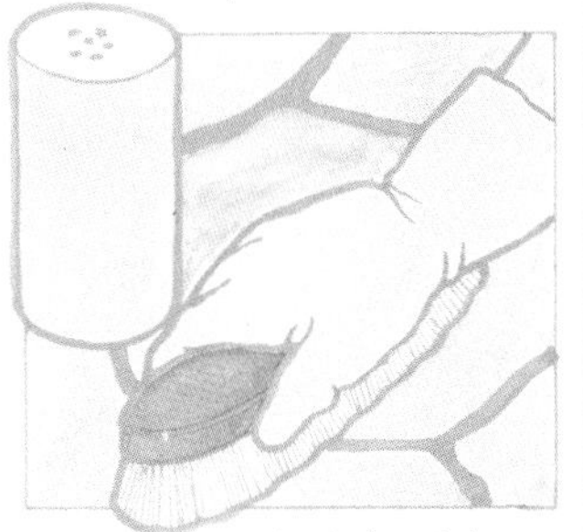

Natural stone looks lovely but scrubbing it clean is heavy work.

Preparation and maintenance

The structure and condition of existing floors or sub-floors can influence your choice. You'll have to check the load-bearing capacity of the sub-floor, for example, if you want to lay heavy tiles. How the floor is constructed will be relevant when you are considering installing new timber boards. Find out if you need to maintain access to underfloor services and, if so, how this can be achieved.

The type of floor will also dictate how much preparation is involved and the degree of disruption installation will cause. Whatever flooring you choose, however, you must have completed major alterations and any messy decorative work before installation.

All floors need regular care to look their best. Check that the maintenance is at an acceptable level for you.

	Outlay	Ease of DIY	Ease of cleaning
Carpet	Fairly cheap to very expensive	Carpet squares – very easy. Fitted carpet – very hard.	Generally quite easy to clean depending on colour
Vegetable fibre matting	Very cheap to quite expensive	Easy.	Doesn't reveal stains. Quite easy tc clean.
Old wood	Very cheap to middle price depending on state of floor	Fairly straightforward.	Doesn't show dirt. Easy to clean when sealed.
New wood	Fairy cheap to expensive depending on wood selected	Possible. Better to have it done professionally.	As with old wood.
Cork	Cheap to middle price	Fairly easy.	As with woods.
Vinyl	Very cheap to quite expensive depending on composition	Sheet vinyl can be tricky. Tiles generally easier.	Dirt-disguising qualities depend or pattern and shade.
Linoleum	Fairly cheap to quite expensive	Again, sheet lino is more difficult than tiles.	Good on all counts. Like vinyl, very easy to clean.
Hardboard	Very cheap to cheap	Easy. Sheets can be cut to shape desired.	Quite easy to clean if sealed. Stains badly if unsealed.
Chipboard	Cheap to fairly cheap	Fairly easy. Tongued and grooved type available.	As with hardboard.
Rubber and synthetic rubber	Quite expensive to expensive	Care must be taken when cutting sheet. Tiles easier.	Dirt tends to collect round pattern reliefs.
Plastic	Matting: Very to fairly cheap Duckboard: Quite expensive	Fairly easy. Comes in clip-together system or roll.	Very practical and easy to clean.
Brick	Fairly cheap to very expensive depending on rarity of type	Possible, but great care must be taken.	Practical and easy. Pale colours show dirt more.
Ceramic	Fairly cheap to very expensive	As with brick.	As with brick. Unglazed tiles need washing often. Absorb oil.
Quarry	Fairly cheap to quite expensive	As with brick. The problem is achieving level surface.	Practical. Need much care in maintainance.
Terrazzo	Expensive to very expensive	Should only be attempted by a professional.	Very practical and easy to care for and clean.
Marble	Expensive to very expensive	As with terrazzo.	Show dirt. Some stains require specialist cleaning.
Slate	Expensive to very expensive	Possible but not easy.	Good in every way.
Stone	Quite expensive to very expensive	Possible with the aid of a helpful supplier.	As with slate.
Concrete	Cheap to middle price	Possible, but heavy work.	Good generally, but reacts badly tc grease and oil.

Durability	General comments
Varies according to quality and amount of use.	Comfortable underfoot and to sit on. Huge range of choice in thickness, pattern, colour, texture, material and cost.
Fine varieties not durable. Coir and sisal very hardwearing.	Many different types of matting are now available. Shades from cream through green to brown.
High.	Handsome and goes with almost anything. Comfortable to stand on and wide variety of stains available.
High.	As with old wood. Both are warm too, if gaps between boards are carefully filled.
High if properly protected.	Comfortable, warm and very attractive, but extremely narrow colour choice.
Very high.	Extremely practical but not often suitable aesthetically for bed- or living-rooms. Wide pattern and colour choice.
High.	More tough looking than vinyl, but similarly limited in room use. Wide range of colours and patterns.
Unsuitable for heavy wear and tear. Not very water-resistant.	Very little choice of colour, but can be painted or stained when sealed.
Should be flooring grade for any durability. Bad water-resistance.	Like hardboard, but perhaps a little warmer in feel. Can also be painted or stained.
High.	Has a sophisticated, modern appeal. Various relief patterns available, but really only one colour.
High.	Looks bright and contemporary, but range is fairly narrow.
Very high.	Hard underfoot and tiring to stand on, but range of colour, texture and thickness improving all the time.
Very high.	Huge range of colours, patterns, shapes and textures. Always looks smart and contributes different atmospheres.
Very high.	Ideal for rustic, cosy, warm-looking effect. Shade pleasing but very little variety.
Very high.	Very smooth and sophisticated in appearance. A fairly wide range of colour and pattern available.
Very high.	Ultra-elegant but cold and tiring to stand on. Depending on quarry of origin, marble ranges from white to black.
Very high.	Dark and natural looking, but no range of shade to speak of. Not quite as cold as marble.
Very high.	Theoretically, stone comes in a variety of shades, but the range is actually rather narrow.
Very high.	Not really suitable for indoor, visible use. Can easily look ugly, but slabs are fine for paths or patios.

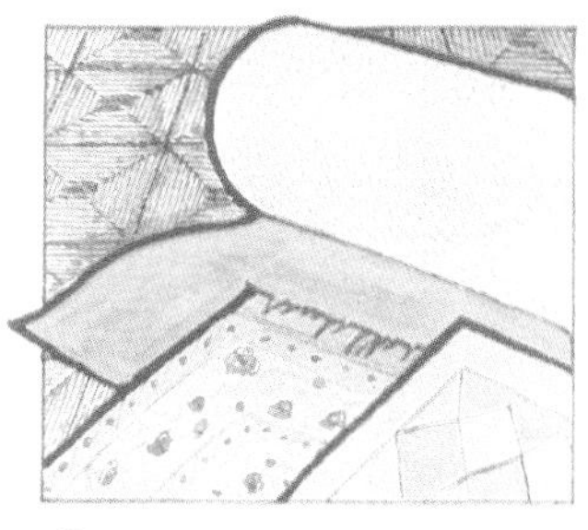

Carpets, rugs and matting

Wooden flooring

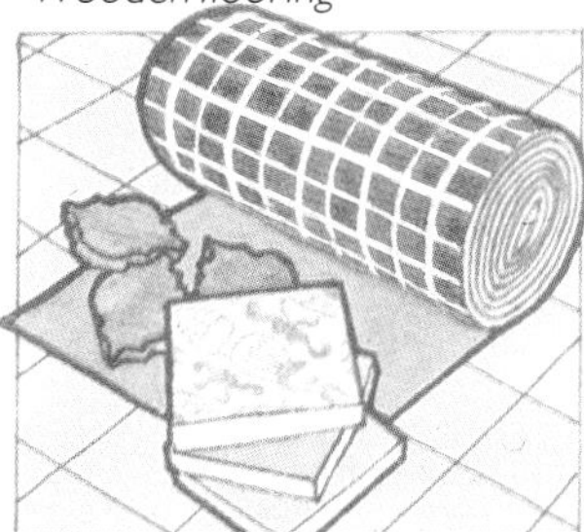

Tiles and sheet flooring

Hard flooring

CARPETS

Warm and comfortable underfoot, available in a wide range of colours, patterns and textures, carpet is deservedly popular. A fitted carpet can do more to enhance and unify a decorative scheme than almost any other type of floor; and modern fibres will stand a high degree of wear.

In terms of durability and appearance, there is almost nothing to choose between *good quality* (well-constructed) carpets of a similar pile weight. The price and performance of a carpet is more likely to relate to the type and amount of fibre used and to the *quality* of construction, rather than the method of manufacture, though tufted carpets, for example, are generally cheaper than woven.

Woven carpets

Fibres available include all-wool, all-synthetic and mixes. The backing and pile are constructed at the same time. Three types of warp yarns are intertwined into the side-to-side frame of weft yarns. *Chain* warps go over and under the weft yarns to bind them; *stuffers* run through the centre of the carpet to fill it out and stengthen it; and *pile* warps form the carpet surface.

WILTON can be plain or patterned. As well as the smooth-cut pile usually associated with Wilton, there is Brussels weave (uncut loop pile), cord carpet (low loop pile) or a mixture of cut and looped pile. The pile warp yarn is continuous and carried underneath the weft, then over a wire which pushes it to a variable height above the backing in a loop and fixes it. These loops can be left uncut, or sliced to produce a thick, smooth pile surface. By mixing cut and uncut loops, sculptured effects can be achieved; by using long wires, 'shag' style Wilton carpets are produced.

To obtain pattern, up to five colours of pile yarns are run through the weave, controlled by a punched card-operated jacquard, which lifts the correct colour from its frame, carrying the others through the backing as in fair isle knitting. All this extra fibre means patterned Wiltons are usually heavy top grade, but very classy plain ones are made with the jacquard process.

Lower quality plain Wiltons are made by the face-to-face

Carpet fibre chart
Dots indicate quality and performance

	Price	Feel and appearance	Ease of cleaning	Dirt performance	Durabi…
Acrylic (Acrilan, Courtelle, Orlon etc.)	Fairly costly	●●●●●	●●●●	●●●●	●●
Cotton	Cheap	●●●●●	●●●●●	●●●	●●
Nylon (Timbrelle, Enkalon, Antron etc.)	Good quality will be costly	●●●	●●●●	●●●●	●●●
Polyester (Terylene, Dacron Trevira etc.)	Cheap	●●●●	●●●●●	●●●	●●●
Polypropylene (Propathene, Fibrite, Olefin etc.)	Cheap	●●●	●●●●●●	●●●●●●	●●●
Sisal	Cheap	●●	●●	●●●	●●●
Viscose rayon	Cheap	●●●●	●●●●●	●●	●
Wool	Costly to very costly	●●●●●●	●●●●●	●●●●●	●●

Wilton cut pile. *This is woven on a loom in loop form, the loops being cut to make the pile.*

Wilton loop. *Known also as Brussels weave, it is woven as cut pile, but the loops are uncut.*

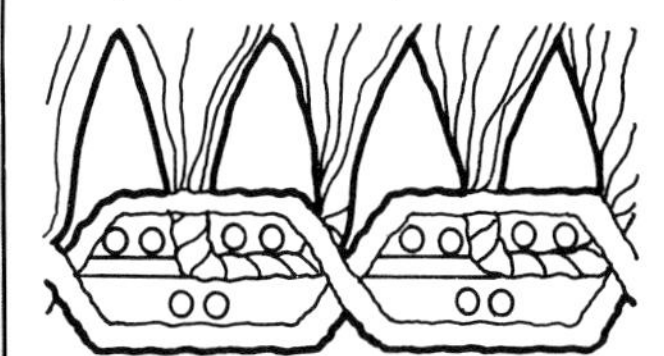

Axminster cut pile. *Each tuft is separately inserted into the backing before being cut.*

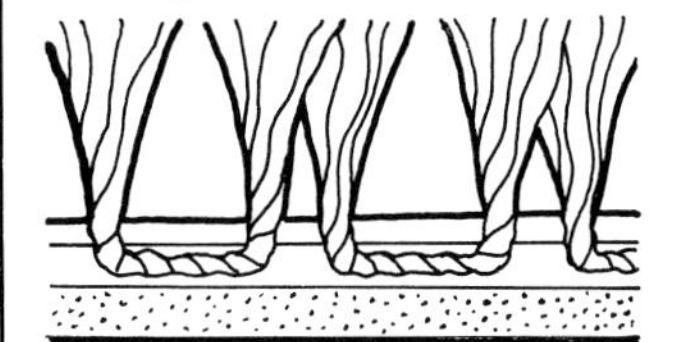

Tufted cut pile. *A fairly new type of carpet, cheaper than Axminster or Wilton.*

Tufted loop. *Basically, the same as the cut pile with the loops uncut.*

method in which one set of pile warp yarns is shared between sets of backing yarns. As the sandwich is made, it is sliced through the centre to separate it into two carpets.

AXMINSTER For gripper Axminsters, each tuft is separately inserted into the backing by grippers and then cut, which means that no unused yarn runs through the carpet and that a large number of colours can be used – usually about eight, sometimes as many as sixteen – all controlled by jacquards.

Less tightly woven, but more highly coloured, are spool Axminsters: the yarn is wound on spools arranged over the loom in as many rows as needed for pattern repeat, and these clank round continuously. Up to thirty to forty colours can be used this way – sometimes with regrettable results! The back of gripper Axminster is ridged and the pattern does not show. The back of spool Axminster is smooth, usually with the pattern showing through.

Tufted carpets

Introduced about thirty years ago, tufted carpets are made by inserting pile yarn (synthetic or wool), threaded through a row of needles poised above the machine, into the prepared backing. Once in, the hooks hold back enough yarn to make a loop while the needles are pulled out and the process repeated. For cut pile, the loops are sliced as they are made. The carpet backing is then coated with adhesive to secure the tufts in position, and usually another backing is applied, too. In recent years it has become possible to achieve flecked effects, and even simple patterns. Most patterned carpets in the United States are printed tufteds.

Non-woven carpets

These carpets can be *needlepunched* and fixed with a coating of adhesive. Or they can be *bonded* – by adhesive which fixes pile fibres on the pre-manufactured backing, in ridges for a cord effect. This can be done simultaneously on to two sets of backing which are then sliced through the centre to make two smooth-faced carpets at once. Or they can be electrostatically *flocked* in which (usually nylon) short fibres are electrostatically attracted to the backing.

Colour, texture and pattern

A wide expanse of a single-colour carpet can dictate the whole atmosphere of a room. This brilliant green carpet blends with the green seating, and combined with white walls and table, the effect is cool, fresh and stimulating.

Colour

A smooth, plain carpet all in one colour is in many ways the best background for a room designed to show off furniture, paintings, or similar features, since the carpet won't argue or compete. Choice of colour is, however, all-important. Muted shades of blue, green, yellow and pink, for example, are classic drawing-room colours; muted grey and lilac newer arrivals. Bright strong shades would suit a room decorated in primaries or in white. Oatmeals, beiges and browns can be comfortable, neutral backgrounds but, if everything else displays an equal restraint, they will be dreary and dull. Black carpet is striking, but shows up lint, fluff and dust very readily; white carpet spells luxury but shows dirty marks (on the other hand, it inspires people to clean it more quickly). Dark shades make a room look smaller; pale ones increase the impression of spaciousness. The amount of light a room receives is relevant too – a very dark carpet in a dark room will look dead unless there's a deliberate attempt to design a room for dark living, with rich colours and clever lighting.

Texture

Textured carpet can be very hard-wearing, particularly if combined with a good underlay, and also adds visual interest to an interior. The raised surface of a cord carpet, for example, smartens up a room and provides a subtle, but discernible, variation without the distraction of a pattern. While cord is a workhorse surface, other varieties of sculptured carpet are more luxurious, and look particularly good in shades of cream, oatmeal and white.

Another type of textured effect can be achieved with colour alone as in 'berber' carpets, loop pile carpets where the colour varies in the loop. The overall effect is a pleasant speckle centred on a main shade. Most berbers are sand, brown, beige and cream. For those who want the effect of a single colour but don't like the way dirt shows up immediately, or who perhaps find a single colour too flat and monolithic without wanting to grapple with pattern, berbers are the answer: they disguise dirt, don't 'track', have a certain depth and are usually hard wearing.

Left: A subtle, chequered texture adds interest to a neutral carpet used to give an effect of space.

Below: The stylish geometric pattern in this pale carpet sets the classical tone for an elegant drawing room, simply furnished.

Pattern

In design-conscious circles, patterned carpets, except for rugs, have been dismissed, if not despised, for so long that the law of averages dictates a revival in some form is due. The trouble is that large-scale floral patterns jostle with patterned upholstery and walls and give the eye no rest. But William Morris, the great nineteenth-century British innovator in interior design, created patterns for carpets that fitted in very well with decorative schemes and many of his ideas are being revived today. Geometric or stylized patterns in rich or pastel colours can also look very good.

In general, it is the size of the carpet that is the critical factor. Pattern is often more acceptable on smaller areas, such as rugs. Or try a restrained pattern, such as dots, dashes and daisies, tiny squares or small diamonds. A carpet with a confetti of pattern like these, particularly if it has muted tones or just two or three colours, will give a contemporary feel and can even work throughout a home. But if you do install a heavier duty carpet in the hallway, for example, make sure the two do not clash.

A bold, abstract design in dramatic colours is not for the faint-hearted but, coupled with plain painted walls and restrained furniture, can add vitality to an interior. If you've a sure touch, mixing strong patterns and decisive furniture can look tremendous: the epitome of decorator style. Patterned carpet can successfully emphasize a theme – a jungle print would carry on the lush effect created by trailing ferns and eastern bric-à-brac. Although horizontal stripes make narrow spaces seem wider, big patterns generally diminish a room. They also disguise dirt, which is why rich and fussy patterns are found in public places. Do you really want your living room to look like a hotel lobby?

Laying carpet

Measuring for carpets

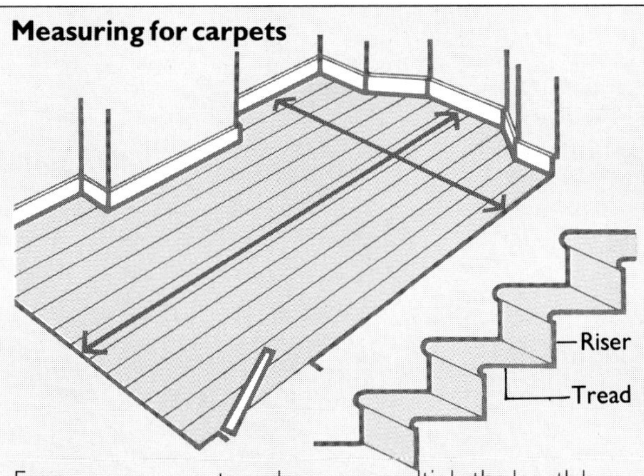

For a square or rectangular room, multiply the length by the width. Few rooms are absolutely square, so take the longest wall in each direction if there is a difference. This calculation gives you the number of square feet, yards or metres according to your measuring system. To convert square feet into square yards, divide the number of square feet by nine. For strip carpet priced by the linear yard, decide which way the carpet is to run. Ideally, it should be laid at right angles to the main source of natural light, to minimize the joins visually. It is also a good idea to lay it in the same direction as the room's traffic movement. In practical terms, lay it whichever way requires the fewest strips to be cut. Measure the width (or length) of the room in inches, then divide by 26 or 36 (according to the width of the carpet) to arrive at the number of strips needed. Multiply this number by the yard length (or width) of the room, for the number of linear yards needed. For patterned carpet, allow one matching repeat per strip.

For stairs, measure each stair separately. Calculate the length of carpet by running string from top to bottom of the staircase, down the risers and over the treads. Add an extra 60cm (24in) so the carpet can be moved up and down to even out the wear and tear.

On the whole, carpet laying is a job for the professional. If you want to attempt it yourself remember you'll be dealing with unwieldy, heavy rolls of material which may be hard to cut and must be stretched to fit – a tricky process. The correct equipment is important too, and not cheap. However, some manufacturers produce kits designed for the do-it-yourself fitter and other manufacturers supply instruction manuals.

Ordering a carpet

It's a good idea to measure the space to be carpeted before shopping around, so you can compare carpet prices and estimates accurately. This will just be a rough guide – the supplier or carpet fitter will take exact measurements to prevent any subsequent disputes.

Work out the rough price by multiplying your square metre (yard) figure by the price per square metre (yard) and adding on the amount for underlay, fitting and any purchase (value added) tax. Another approach is to work backwards from a budget limit to find out how much you can afford per square metre (yard) laid.

With fitted carpet you probably won't be able to avoid either wastage or seams and minimizing both at the same time can be tricky: the art of laying carpet involves resolving these sometimes contradictory demands. Making a plan of the spaces to be carpeted will help.

Before the floor is ordered and laid, assuming you aren't laying the carpet yourself, the carpet suppliers should give you an estimate. This should state: the type and amount of carpet (by manufacturer's name, code, with width and colour specified) and a similarly exact description of any underlay; the rooms/area to be covered; the cost of the carpet, underlay, fitting and tax, separately detailed.

The fitters should give you a date and time for delivery and fitting and specify whether their charge includes moving furniture, removing doors and so on. Some fitters will supply you with a diagram of where the seams will go, but you should at least discuss it beforehand. This important point relates to the width of carpet you have chosen: it would be foolish to take a narrow width for a large space. Further, with velour or

Carpet tiles

These are small squares of sealed-edged carpet available in several sizes, materials and colours, both plain and patterned. Most common are 40cm (15½in) and 50cm (19½in) squares, but 30cm (12in) and 45cm (18in) squares are also available. Carpet tiles are made in a wide range of fibres, including wool, other animal hair, synthetic fibres and mixes. Backings can be of PVC, polypropylene, natural and synthetic rubber and bitumen impregnated felt. Whichever sort you choose, it should possess guaranted dimensional stability. Carpet tiles laid in a single colour will resemble broadloom carpet; chequerboard effects can be created by using two colours, and other, more complicated patterns are possible. They can be stuck down, or loose laid for easy replacement, and are ideal for dining rooms and children's rooms, where dirt, damage and general wear and tear are likely. In very worn-out areas, though, a single, new carpet tile can be as conspicuous as a damaged one; moving the tiles around regularly is often a better solution. Carpet tiles allow access to under-floor services.

patterned pile, you can't just add in extra pieces – the pattern repeat and pile direction must match up. The method of fitting dictates cost, too – sewn seams cost more than stuck seams, for example.

Types of underlay

It really is worth spending money on a good quality underlay – even if the carpet is cheap, underlay may extend its life as well as improving the way it feels underfoot. Foam-backed carpet does not need underlay, but should be laid on felt paper (very cheap) to prevent the foam sticking to the floor.

Underlay absorbs pressure on the carpet, lessens wear, cushions the carpet from unevenness in the floor, prevents dirt from working up from the floorboards and protects the carpet from rot. It insulates for heat and from noise and is pleasant to walk on. Beware, however, of buying a too-thick underlay: heavy furniture may sink in and pull the carpet out of shape; also doors will have to be trimmed.

FELT underlay is usually made of jute or animal hair, or a mixture of the two. Jute felt, while cheaper, can flatten, harden or break down. Hair felt is generally stronger, more resilient and durable. A mixture is reasonably strong and cheap. The weight and thickness of this underlay should be at least 1628g per square metre (48oz per square yard) in a heavily used room, and underlay should have been moth-proofed by the manufacturer.

FOAM OR RUBBER underlay is made from natural or synthetic materials. Both types are fairly springy and don't shed fibres, but they can rot in damp, may perish on an overheated sub-floor, and are unsuitable for heavily seamed areas or stairs. They are best with a hessian or mesh nylon/polypropylene backing fabric or a textured surface. To test quality, rub a foam underlay firmly with your thumb – it should not crumble at all. Reject it if it does.

RUBBERIZED FELT theoretically combines the advantages of both rubber and felt.

BONDED UNDERLAY is similarly made from wool, synthetic fibres and latex rubber.

Preparation

As always, lay only on a sound, dry surface. Remove all old tacks and nails or hammer nails in flat. Fix loose floorboards and correct severe unevenness in the floor. Cover floors in bad condition with hardboard. Do not lay carpet direct on thermoplastic tiles; moisture cannot escape and will eventually damage the carpet.

Laying woven carpet

Woven carpet is best laid by fitting it to gripper rods. These are quite expensive but can be re-used, and the laid carpet looks smooth to the edges. The gripper is a strip of light wood 2.5cm (1in) or so wide, with pins protruding at an angle; is nailed or stuck around the edges of the floor with the pins facing the wall. The carpet is stretched, hooked on the pins, and tacked neatly down into the gap. Where the carpet does not run up to a wall a binder bar is used: the carpet is pulled on to the pins and then covered by a curved metal strip. Double binder bars are used where two carpets meet.

To fit pre-nailed gripper rods, use a hammer similar to a carpet fitter's hammer, with a small striking face on a heavy head so you avoid striking and blunting the gripping points. To follow curved recesses in the room, cut the strip into very short lengths.

TACKING is a traditional, cheaper method of laying which involves stretching the carpet, turning its edges in and tacking them down every 7.5 to 10cm (3 to 4in) with 2cm (¾in) tacks, 2.5cm (1in) where the carpet is very thick. But the indentations can look untidy and will collect dirt.

RING AND PIN is a type of laying that enables the carpet to

Laying woven carpet

1. Nail gripper strips around room. Leave gap between strips and skirting boards.

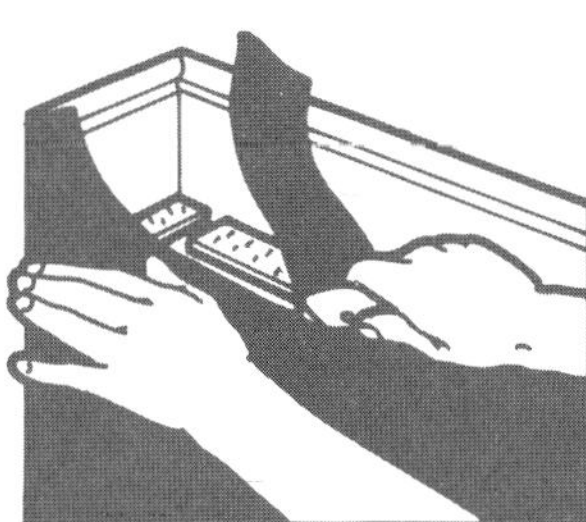

2. Fix underlay up to gripper strips. Secure with tacks or strong staples.

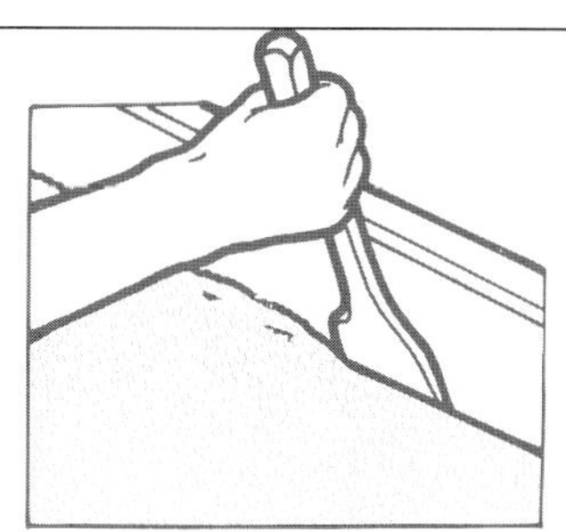

3. Fix carpet over gripper strips. Push down to fill gap so it butts up to the skirting.

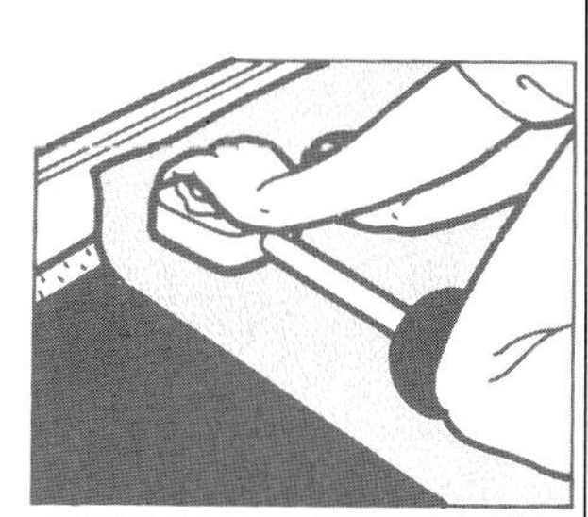

4. With knee kicker, gradually stretch carpet to opposing walls until it hooks gripper.

Laying foam-backed carpet

1. Line floor with felt paper. Push carpet firmly into position using a board.

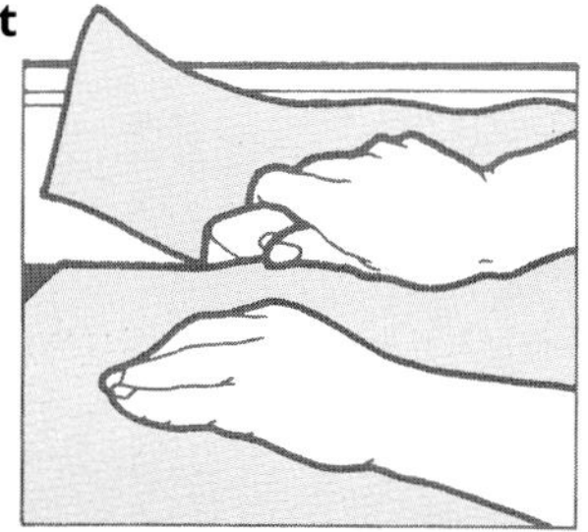

2. Cut carpet with a small overlap. Allow it to settle for a week, then trim it to fit.

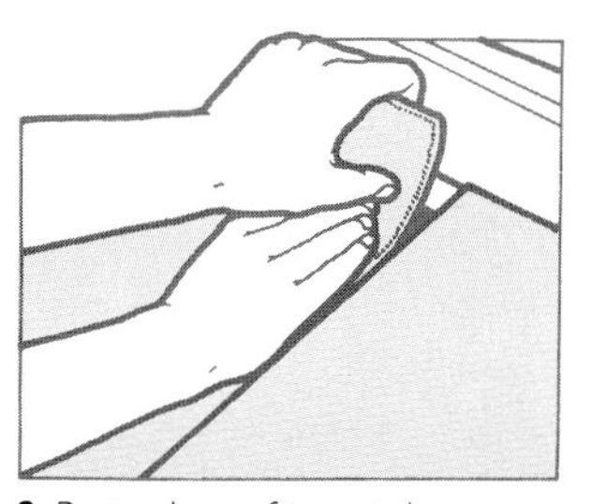

3. Butt edges of two strips as close together as possible, using double-sided tape.

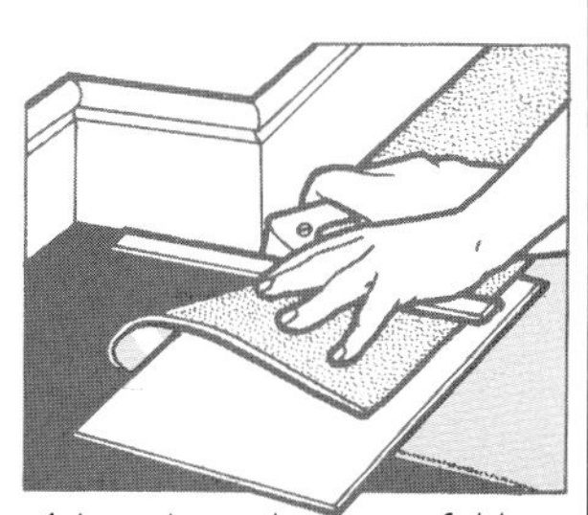

4. In awkward corners, fold carpet back over board. With ruler, extend wall line and cut.

A good-quality fitted carpet is a sensible investment for stairs; it looks attractive and muffles sound. For safety and long-lasting wear, fit the carpet securely. To use gripper strips, fit one on each riser and one on each tread, with the pins facing the angle between. Make the gap between the two strips twice the thickness of the carpet. Once the strips are in position, fit the tread underlay mats between the grippers. Tack, stick or staple each mat securely, then fit the carpet starting from the bottom of the stairs and working upwards. Stretch the carpet and hook it on to each gripper as you move up the steps. Stair rods, fitted across the carpet where each riser and tread meet, are another alternative.

A runner can be a less expensive alternative to a fitted carpet, but depends for its beauty on the exposed staircase being well finished. It, too, needs secure fixing.

be taken up for cleaning but is expensive and must be done professionally. Rings are sewn into the carpet's backing and then fixed over pins or a gripper strip in the floor.

FIXING UNDERLAY The underlay should stop just short of the gripper rods and (unless it's foam) be fixed with tacks, although staples, if you can obtain a stapling hammer, are better. Cut the underlay slightly oversize, fix it, then trim to fit along the guiding inner edge of the gripper strip. On concrete or similar floors stick it around the edges of the room.

FIXING CARPET Begin in a corner. Let the carpet ride up the wall a little and smooth it on to gripping pins by hand. Using a fitting hammer or similar tool, push it down on to the points and draw it along the wall, forcing it into the gulley. Do this for the two walls that meet in the starting corner.

Then stretch and hook carpet on to the gripper strip on opposite walls. If you can borrow, hire or even make a carpet stretcher do so, but practise first – you could hurt yourself and chew up the carpet. The 'teeth' of this device dig into the carpet and a push with the knee stretches the carpet a little, a process which is repeated until the carpet stretches enough to fit over the gripper strip, after which it contracts and stays in place. If you don't have a stretcher, push the carpet along with your feet, making little jumps. Knock in tacks lightly to help hold the surplus you've pushed along.

For alcoves, cut carpet long enough to reach to the end, then make a cut on both sides at right angles so a tongue of carpet falls into the alcove. The same principles apply to tufted and non-woven carpets.

Laying foam-backed carpet

Foam-backed carpet does not need to be stretched. Cut the carpet just slightly larger than the room – 2.5cm (1in) or so extra at each wall. Trim to fit. To join at seams, use wide carpet seaming tape.

Foam-backed carpet, needlefelt and carpet tiles can be stuck down all over with adhesive, or loose-laid and secured with double-sided tape. Foam-backed carpets should never be laid directly on to polished, varnished or vinyl floors or they will stick permanently – lay paper felt first.

RUGS

Rugs lift your decorative scheme out of the ordinary, adding colour, accent and life. Easy to move and available in a wide price range to suit every pocket, they are the ultimate in flexibility.

The quality of a hand-made knotted rug has to do with density – the more knots the better. Check by examining the back to see how fine the knots are and how many there are per square centimetre (inch) – but some rugs with fewer knots may be of a superior grade of wool. Patterns should be clear, balanced – even luminous. (Printed reproduction rugs will look blurred and muddy.)

Aside from the traditional types, there is a wide range of contemporary rugs produced both by the carpet industry and by individual craft makers. Manufactured versions come in all price brackets and designs, including some very passable reproductions. But for a really exciting alternative, there is nothing to beat a rug commissioned from a designer-craftsman – either to hang on the wall or as the centrepiece of your living room.

For a cheerful, cosy look, rag, braided and hooked rugs still have their own thriving traditions. Antiques now fetch high prices but homemade versions are easy to do with scraps of suede, string, silk, cotton or wool.

1. Chinese Thick, rich rugs, mainly in silk or wool. Black, soft yellow, pink, peach, apricots and blues are favourite colours. (Illustrated is an antique Pao Tao, in muted shades.) Slightly sculpted look; motifs include dragons, lotus blossom. Patterns often have meanings – the butterfly is a symbol of prosperity, peonies stand for love and affection. Expensive.

2. Numdah Pretty, soft and colourful, but usually made of felt and won't stand up to heavy wear. Often small and circular or oval, with naive designs on a cream or off-white ground. Extremely cheap to cheap.

3. and **8. Persian** Originated in central Asia, still the centre of a rug-producing area, though some types rather bastardized. Almost always rectangular or in long runners, knotted in wool or silk (or a mix) on to a woven base. Usually richly coloured (reds and blues are favourite grounds) with stylized motifs. 3 is made by country nomads, 8 in a small factory. Quite expensive to extremely expensive (particularly for rare, very beautiful or antique pieces).

4. and **7. Dhurry** Indian flat-woven, hand-woven cotton rugs; some expensive types may include wool or silk. Available in fantastic range of colours from pastel to bright; usually geometric designs, stripes, simple borders. They can be reversed because they have no pile. Very cheap to medium-expensive.

5. Tibetan The rug making skills and styles of Tibet are still continued today, often by Tibetan exiles working in neighbouring countries. (The double-bordered rug illustrated here, known in the trade as a Kangri, comes from Nepal.) Can be expensive.

6. Turkish Usually from the Anatolian region. Often prayer rugs, so have pointed prayer arch, or pointed shapes, at one end, sometimes flanked by pillars – but each region has its own distinctive patterns. (Illustrated is a washed Turkish.) Expensive.

9. Flokati Greek shaggy wool pile in rectangular sizes from small to large, often in cream and grey. Very cheap to moderately expensive.

10. Kelims Made on a small scale, essentially in Afghan and Turkish versions. Both types have geometric patterns, no pile and are tapestry woven, usually all in wool. The Turkish type (illustrated here) is usually cheaper and brighter, of mixed design, with naive motifs of great charm and often embroidery too. The Afghan version is of tougher, thicker wool, in dull rich colours and plainer patterns, and is subdivided into Baluchi types and those from northwest Afghanistan, woven by Turkoman tribes. Many of these pieces are now collectors' items. Quite expensive to very expensive.

Right: Three similar, but not identical, Indian rugs frame this delightful bed, and are perfectly placed for bare feet on a cold morning.

Below: Underlay is essential to prevent rugs slipping. Many varieties are available; this one is made of webbed, long-lasting slightly sticky material, cut from a roll to fit rug size.

Laying and positioning rugs

Almost all rugs benefit from a soft underlay. This could, in fact, be a fitted carpet, but dhurries, numdahs and other non-dye-fast rugs should be laid on interlining in case the colour rubs off, which it may do particularly if the rug becomes wet. It is also important to secure rugs where people might skid and fall.

There are several varieties of useful anti-slip material, often made in a honeycomb of slightly sticky flocked material, which, to some extent, prevent rugs 'creeping' when laid on carpets, or skidding when laid on a hard, shiny floor surface. At the very least, use double-sided adhesive tape.

A strikingly designed or patterned rug must be considered within the context of your overall interior scheme. While it is surprising how even the brightest rug can simply fit in with your furnishings, mistakes can be made. Always check whether the colours will clash with your decor. Don't be afraid to banish other forms of pattern, letting the rug dominate; or to offset vivid rugs with simple white walls; or to lay pattern on pattern. But try and keep the pattern types compatible – a Scandinavian contemporary rug just isn't going to look right next to a faded runner. Do not place a valuable rug, especially an antique one, in areas where it will suffer traffic: they were never designed for western living.

Care of rugs

Rugs benefit from being moved regularly, to even out wear and fading. It is also important to deal with spills and stains as

Stain removal for carpets and rugs

Alcohol	Blot, wash, surgical spirit if necessary.
Burn	Clip off burned fibres, wash.
Coffee	Blot, wash, dry clean if necessary.
Fruit juice	Blot, wash, methylated spirits if necessary.
Grass	Methylated spirits, wash.
Grease, oil	Blot, dry clean, wash, repeat if necessary.
Ice cream	Scrape off, blot, wash.
Ink	Blot, wash or methylated spirits.
Milk	Blot, wash, dry clean if necessary.
Paint	Scrape, dry clean if oil based, or wash.
Shoe polish	Scrape off, dry clean, methylated spirits.
Tea	Blot, wash, borax solution if dried stain.
Urine	Blot, wash with dash of antiseptic.
Vomit	Scrape off, wash with dash of antiseptic.
Wax	Scrape off, iron lightly over tissue paper, methylated spirits if necessary.
Wine	Dilute with soda siphon, blot, wash.

BLOT Absorb liquid rather than rubbing it into the carpet. Use a well wrung-out, clean damp rag, dabbing it in quick, short strokes. If you use salt to absorb wine, vacuum up immediately afterwards, as it can affect dyes.

WASH Try warm water with a dab of white vinegar first, then warm water with a tablespoon of salt, then a weak, warm detergent solution and, if all else fails, a light solution of carpet shampoo with a dash of white vinegar. Sponge and blot the area, using the shampoo foam, then rinse. Leave fibre pointing in the correct direction and lift the carpet clear of the floor to dry. Water-based, starchy, and sugar-based food and drink stains should be blotted, then washed.

DRY CLEAN Ideally, put an absorbent pad under the carpet or rug. Test first, then dab with dry cleaning solution, from outside of stain towards centre, changing the pad frequently. Do not use on foam-backed carpet unless necessary; do not allow fluid to touch backing.

they occur and not to use strong detergent solutions, unless the manufacturer specifically permits it. Beat the rug regularly with a flat-faced carpet beater in preference to vacuum cleaning. Brush away loose fluff by hand. Otherwise, vacuum clean on both sides, going with the pile, not against it. Use a small cylindrical cleaner with a nozzle like those used for car interiors rather than an upright vacuum cleaner. Clean the place where the rug lies before you put it back.

Some rugs can be washed gently with a good-quality carpet shampoo that dries to a powder and is then vacuumed. A little white vinegar may brighten colours. Do not overwet the rug, do not hang it up to dry, do not use hot water, do not 'steam' clean, scrub or immerse – and make sure beforehand that washing will not damage the rug.

Replace broken fringes with matching thread; blanket- or circular-stitch damaged edge binding and avoid re-piling if possible as the success of this can vary. But it is quite possible to re-pile or patch in a way that just makes the rug that much more homely; or, with expert help, to achieve an 'invisible' mend – though not, obviously, on antique rugs.

Care of carpets

New carpets tend to fluff: for the first few weeks clean lightly with a brush. Thereafter, vacuum at least once a week so dirt doesn't become embedded at the base of the pile where it can rub and cut fibres loose. Carpet sweepers are not able to reach the roots in this way, but may be handy for picking up surface mess. If a few tufts jut out, don't pull, just snip to the level of the pile. When you vacuum, make sure the bag is no more than half full or the suction will be inefficient. Vacuum slowly, to give the machine time to loosen and suck out dirt.

If you shampoo carpets, follow instructions exactly. Professional spray extraction cleaning (often mistakenly called 'steam cleaning'), which involves injecting high pressure jets of water and detergent into the pile with a powerful wet vacuum cleaner, is best left to professionals.

Remove stains right away – have a stain removal checklist handy, so you know how to deal with the most typical, and keep the right materials handy, too.

MATTING

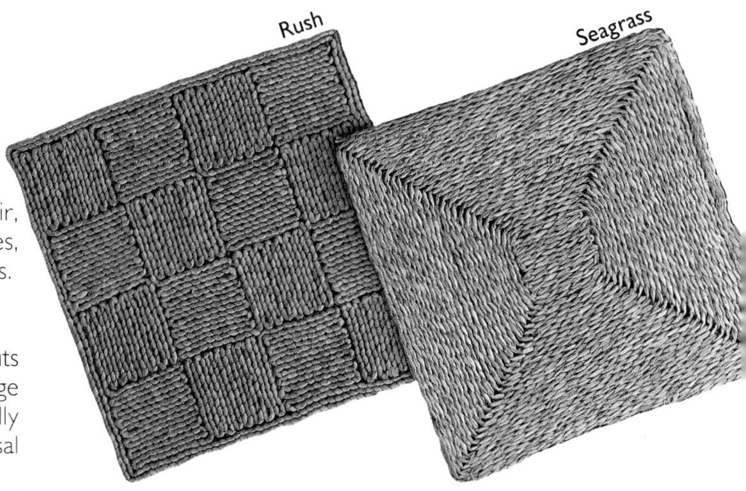

As wall-to-wall flooring or cheerful scatter mats, sisal, coir, rush, seagrass and maize, as well as synthetic plastic types, make cheap, easy alternatives to more traditional surfaces.

Sisal and coir

Both cheap, hard-wearing fibres, coir comes from coconuts and sisal from the leaves of the agave plant. Colours range from pale brown to dark brown, though coir most usually comes in a honey gold. Now it is also possible to dye sisal many beautiful shades, although not all are fast.

Both fibres can be uncomfortable, and may crumble and shed fibres and dust. Better varieties are latex- or vinyl-backed to prevent dust and dirt falling through to the floor beneath and make for greater durability. Good for stairs (but ensure the surface doesn't become hard and slippery), corridors and as a base throughout the house, but avoid using in the kitchen or under dining tables because of food spills.

LAYING After cutting slightly oversize, allow to acclimatize to room for twenty-four hours. For wall-to-wall laying, pick one of the broadloom types, stitch narrower lengths together, or butt edges firmly together, using double-sided tape underneath. You can loose-lay but it is better to stick down using double-sided tape or carpet adhesive. Tacking down edges may cause slight bumping and troughing.

Rush, seagrass, maize

Natural materials of varying fineness of texture, these vegetable fibres have plenty of cool charm. Maize is the finest and palest in colour. Woven squares can be sewn together into mats using fine twine, and there are many attractive weaves. Don't expect them to take heavy wear; they may be quickly shredded if there are cats in the family.

LAYING Do not need underlay. Also look good scattered over other flooring.

Plastic

Woven plastic matting comes in bright colours and has a jolly, cheerful appeal. It is very cheap and fairly durable. Avoid direct heat, strong chemicals and abrasive cleaners.

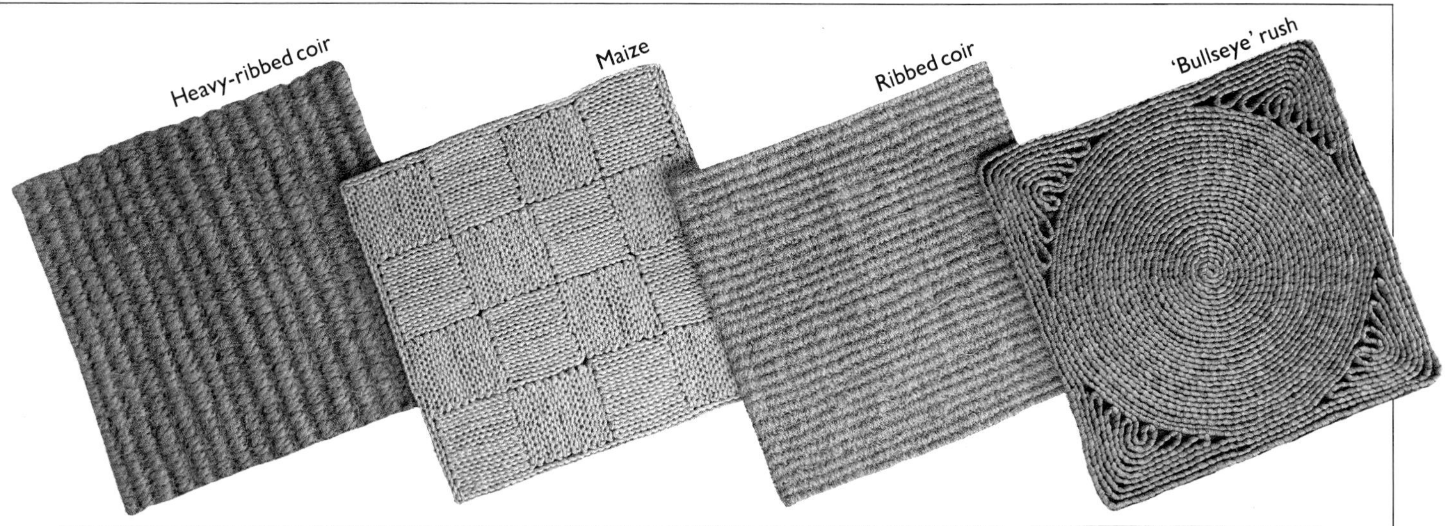

Care of matting

SISAL AND COIR

Vacuum clean. Loose matting can be taken up and beaten, and scrubbed with soapless detergent. Clean underneath now and then if you can.

RUSH, SEA-GRASS, MAIZE

Deteriorate if too dry, so moisten occasionally with a plant sprayer. Lift and sweep beneath and snip off flaking ends. Bind edges if they start to fray.

Far left: The sea-grass matting squares add definition to the sleeping area and provide the visual and physical warmth lacking in the ceramic tiles.

Left: Practical and hard-wearing coir matting is ideal for a garden room. Its natural colour and material are sympathetic to the surrounding greenery, and the subtle diamond weave adds a sophisticated finishing touch.

OLD WOOD

A wooden floor, treated in one of the many ways that costs little but effort, should be durable and economical as well as attractive. Rugs can be added as you wish, which makes for flexibility. While you shouldn't underestimate the amount of time and work that may be needed to execute a finish properly, the sense of creativity can be immensely satisfying. When choosing a finish, consider the room as a whole: the cleverest treatment may only look fussy if everything else is jostling for attention too.

Types of floor

SUSPENDED WOODEN PLANK FLOORS consist of planks supported on joists, support beams which run at right-angles to the planks. Assuming that the joists, and the timber wall plates that support them, are free of rot and damp (which should be checked by a surveyor), the state of the planks will dictate your next move.

Planks come in many widths and generally in cheap, strong wood such as pine or deal. There is a great difference between wide, old solid oak planks and worn, splintery, pine boards; most floors are somewhere in between.

SOLID WOOD FLOORS consist of hardwood parquet, wood strips, tiles or blocks laid on solid floors, usually concrete over hardcore. In upper storey rooms, however, the floor will probably be a suspended one. Solid floors are very durable, but much depends on whether the wood is solid or veneered, and, if veneered, on how thick the skin is, and on the type of base material. Veneered plywood, for example, is not suitable for machine stripping.

Preparation

MAKING GOOD Take up existing floor covering, if any, and inspect the boards carefully. Look out for springy or saggy areas, squeaking boards, wide gaps between boards, flaking and badly pitted surfaces.

If boards are springy or saggy the joists may be at fault; lift the boards to inspect. If they are in a bad condition, seriously damaged or rotten, call in a builder to replace or repair. If they squeak, the boards may not be hammered down

Making good

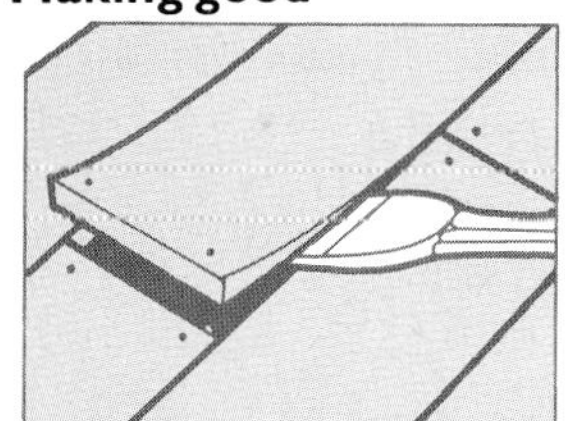

If you need to replace a floorboard, use a special chisel to lever it up.

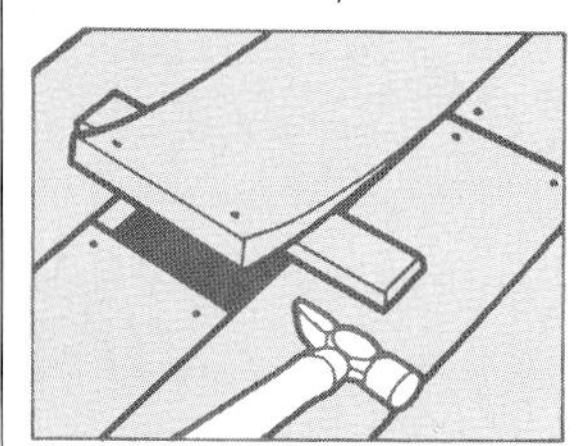

Use a block of wood to prevent the board springing back while you lever it up.

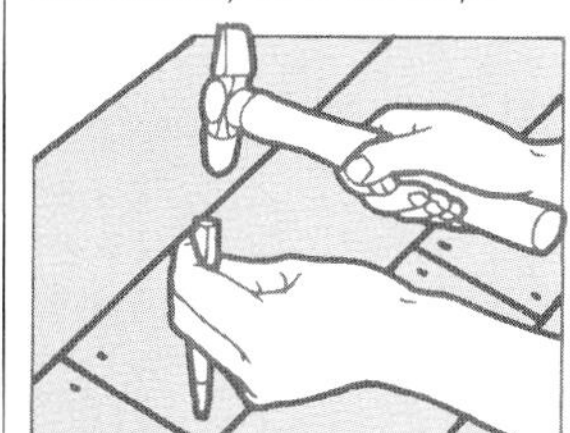

Always sink the head of the nail below the surface of the board, using a nail punch.

Use wood fillets to fill gaps between boards. Or for tiny gaps try papier maché or wood filler.

properly: fix them firmly on the joists with nails, punching them down so that a small space is left above the nail head.

If there are many wide gaps between boards, you would be better off covering the floor with hardboard, but if the gaps are not too serious, fill with wood fillets. Cut matching wood in wedge shapes to fit tightly, glue the edges and tap down smartly into the gaps. Plane until all is level. Smaller gaps can be filled with a homemade papier maché (wet, drained newspaper pulp mixed with glue size); or with a proprietary woodwork filler; or, if you plan to polyurethane the floor, with a polyurethane filler. These filling materials may 'take' stains, varnishes and so on in a slightly different way to the surrounding boards, though you can buy stopping in a range of tints or try to match it yourself using a drop or two of stain.

CLEANING There are easier ways to clean than sanding. Even if it turns out that you have to sand anyway, cleaning will reveal the real condition of the surface and it may emerge that hand sanding will be enough.

First, punch down any protruding nails at least 3mm (⅛in) below surface, using a nail punch. For a professional finish, fill holes with stopping, and if there is a heavy deposit of wax on the floor, remove it with wire wool and white spirit. Wear gloves and scrub thoroughly, but not too wetly, with very hot water and strong detergent. Wetting the wood like this will raise the grain, and, even if you find there's no real need to sand, the raised grain will need rubbing down once the floor dries. Rub down with medium, then fine, glasspaper in the direction of the grain using long smooth strokes. Particularly if the boards have been protected by a carpet or lino you may find they come up very well or with just a few stains and paint splashes which you can deal with individually.

Part of the charm of these old wood floorboards lies in their imperfections – the variations in colour, odd splits and knotting. As with furniture the finish of an old wooden floor is a kind of patina that develops over many years.

Stripping, sanding and sealing

Power sanding

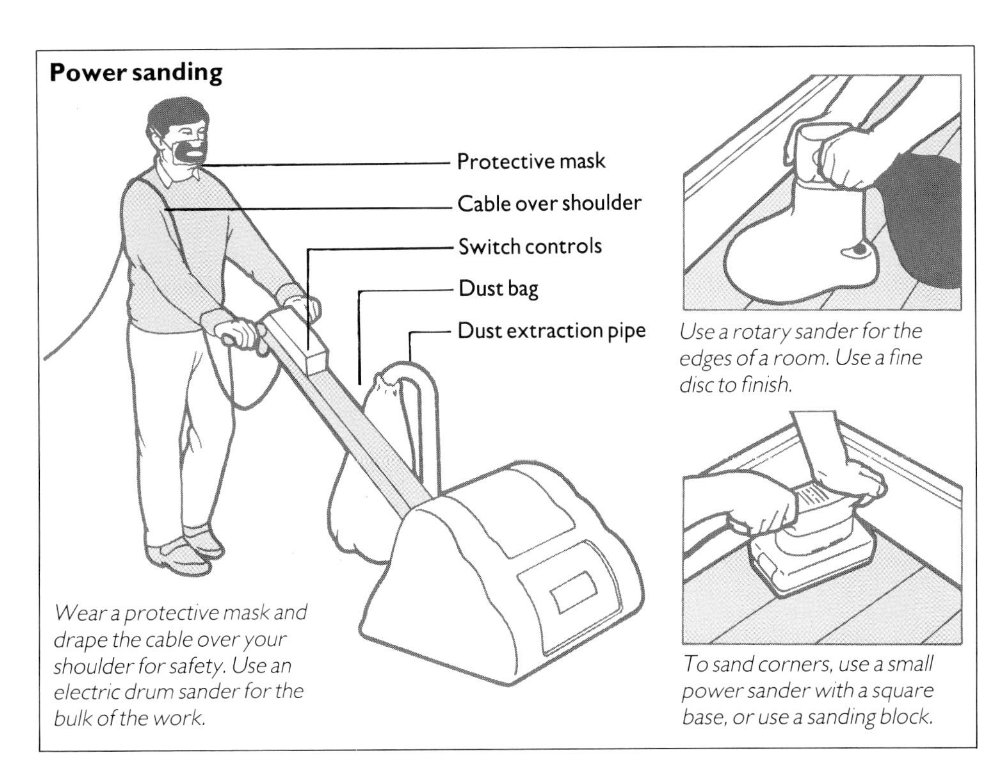

Wear a protective mask and drape the cable over your shoulder for safety. Use an electric drum sander for the bulk of the work.

Use a rotary sander for the edges of a room. Use a fine disc to finish.

To sand corners, use a small power sander with a square base, or use a sanding block.

Stripping

For small areas of floor or stair treads that have been painted, try a proprietary chemical stripper or hot air stripper, and follow the directions exactly. Work quickly, scraping off the softened paint before giving an all-over scrub with white spirit. You could use a blow lamp if you are very skilled, but there's a great danger of burning or scorching the wood badly. If the boards have been treated with a tenacious varnish (or a stain) don't try to clean it off: it may sink in and make matters worse. Sanding is the solution here.

Hand sanding

If the floor displays only a few patches of stubborn stain after cleaning, use coarse, medium and fine glasspaper, in that order, wrapped round a sanding block, rubbing in the direction of the grain. When sanding parquet, sand twice, the second time at right angles to the first. This is because parquet is laid in opposite directions. Give a final, thorough sanding with fine sandpaper.

Power sanding

Boards that are in a truly poor condition, badly stained and uneven, will need power sanding. If there are several floors or a very large area to do at once, power sanding is also a sensible option. If you're dealing with a wood strip or parquet floor, pull up a section to check whether, if it's veneer, the veneer is thick enough to cope with sanding. Glue down any loose pieces with appropriate adhesive.

Diagonally laid flooring plays upon perspective to create an illusion of additional space, and helps integrate traditional wood floorboards into a starkly contemporary setting.

You'll need two sanders: a large upright one with a dust collector, and an edging sander. Wear a face mask and goggles. Using a medium grit belt (or coarse grade if the deposits are thick) on the larger machine, sand at forty-five degrees to the board, then again in the opposite direction (still at forty-five degrees). After this, sand in the direction of the grain, moving progressively to fine sandpaper. If the deposits are light, you need go in the direction of the grain only, but this can create furrows if you do it for too long. Start each sanding run a few centimetres (inches) away from the last to avoid making start marks. Sweep up the dust that settles with a slightly damp cloth. (Do not burn this, or the sawdust that you collect, or you may create a flash fire.)

Do not allow the machine to run on the spot as it will chew grooves in the boards: tilt it up at the beginning and end of rows. Remember that softwood floors will be sanded away quicker than hardwood, but don't be tempted to 'deep sand' stubborn spots: you may end up with potholes. Go over frequently and lightly, finishing with a fine sanding sheet in the direction of the grain. But before the last run with the big sander, deal with the margins of the room with the edging sander, again going in line with the grain. Give a final clean with white spirit: dust must be removed before further treatment.

Sealing

Soft, clean boards must be protected. If you want to decorate or colour them, keep them covered until you have finished and are ready to seal the decoration in. Otherwise, seal immediately. Some traditional wood treatments – oiling, waxing and coating with so-called button polish (shellac) – are not recommended by timber experts for use on floors.

There is no such thing as an ideal seal, but several are equally useful. Oleo-resinous seals, sometimes called clear varnishes, are the result of a reaction between a resin (often a phenolic) and an oil (often tung oil). They are very hard wearing and easy to apply. The almost identical, widely available 'polyurethanes' in a single can are basically oleo-resinous with a small amount of urethane resin added. These two, and most resin solution types, are no more noxious than gloss paints, but some other seals may produce nasty fumes or react badly on skin. Wear protective clothing, a face mask, have proper ventilation, obey the manufacturer's instructions and avoid contact with flame – all seals are flammable.

Urea formaldehydes are lacquers that either have to be catalyzed into effectiveness by separately stored 'acid' or come as one-can 'self-curing' mixtures. They are hard wearing and good for pale woods such as maple since they dry to a colourless film. Other two-can seals are epoxy resin and a heavier duty polyurethane, both durable but slow drying.

Apply seals with brush or pad of non-fluffy cloth, working well into the grain. Do not allow drips or ridges to occur. Seals take a while to cure, so don't subject the floor to heavy wear too soon.

Decorative treatments

Once your floorboards have been cleaned, sanded and repaired – what next? Much depends on your overall intentions and the amount of time you can spare: a floor should fit into a scheme. If you have a lot of space, no rugs and absolutely no money to spare, a painted pattern would make the floor look less stark, or you could paint in your own rug – a well-executed *trompe l'oeil* makes a real talking point. While a simple, glowing, sealed floor is a perfect background in itself, there are so many other alternatives which can be safely preserved under these final tough transparent coats that it might be a pity to miss the chance to experiment. Soft marbling, antiqued brown tracery, sharp black and white squares, boldly painted or bleached boards, thin bright glazes, lustrous wax – all these and more are possibilities.

First, survey your newly exposed floor. New boards standing out starkly from the old or lighter patches of inset fillets will take the finish you apply in a different way to the remainder of the floor. This may not look at all bad; in fact some people rather like the random effect. New boards can be distressed to look like older boards, however, by staining them, slightly roughly sanding them, partially bleaching and staining again.

Painting

If the boards are not in a good enough condition to seal, painting may well be the answer. White boards will be fresh and smart, making a room look clean and spacious. They also look fresh as a border for a runner or rug. Primary colours such as red, yellow or blue make bright, cheerful rooms, while deeper or more muted shades have a more dignified or sophisticated appearance. Hammerite paint has a curious, tough metallic surface which could be interesting in certain applications.

Prepare your surface with great care, making sure it is clean, grease- and dust-free, and smooth. New wood must be checked for knots, which should be coated with knotting, otherwise they will exude resin and blister paint. Buy the best possible paint you can afford: it really is a false economy to do less. And make sure it is the right paint – emulsion won't do,

Stencilling

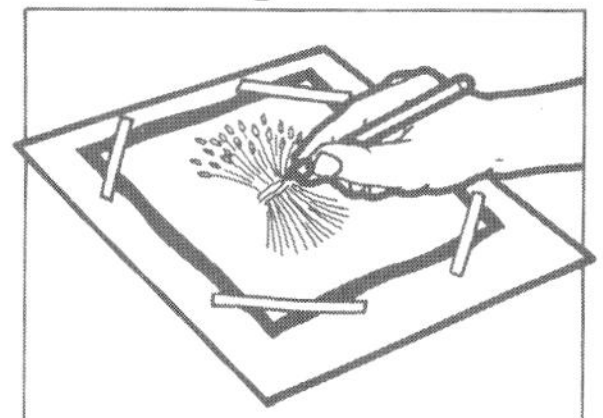

Using carbon paper, transfer the design from the tracing paper on to the stencil board.

Tape down the stencil board; then with little paint on brush, paint through the cut-outs.

Different types of paint

Primer **Pink or white** Use this oil-based sealer on new wood. It prevents top coats from soaking into the wood.
Aluminium This primer should be used with highly resinous woods.
Acrylic This acts as a combined primer and undercoat.

Undercoat **Enamel** This doubles as a primer, but is only available in America.
White With a couple of coats on old wood floors, no primer necessary.
With a dark topcoat, buy a tinted undercoat or tint your own.

Topcoat **Gloss** This oil-based paint dries to a tough, shiny finish. is tougher still when polyurethane added.
Eggshell Oil-based matt finish. White probably best mixing base for any decorating. Takes tints or artists' oils well.
Deck Used for boats etc. Is long-lasting but difficult to apply.

Emulsion This is a water-based, thin paint. It dries to a matt, powdery surface. Needs lashings of protective seal, which can have a slightly yellowing effect.

Left: Many different coloured stains are available. You can choose a colour which bears no resemblance to the natural shade of the wood.

Staining needs plenty of natural light. Artificial light can be misleading. Absolute cleanliness, too, is essential, so remove all dust from the floor before starting work.

To apply the stain, work quickly with a dry, non-fluffy cloth or brush, spreading an even coat over the whole surface, working always in the direction of the grain. Wipe off excess with a clean cloth, still working with the grain. When dry, wipe over again.

Far left: This blue and white chequered pattern is actually meticulous paintwork applied on the wooden flooring of a country kitchen. To take the hard wear of a kitchen, the paint must be extra-carefully sealed.

unless you varnish heavily. After painting, protect the surface of non-gloss paints with seal.

Stencilling

The charm of stencilling lies in its quaint, hand-done quality. Good preparation is essential. This means making or buying a good quality stencil, and using the right equipment. Study common stencil designs in history books, and look at those commercially available, before making your own: it is important to have enough solid space between the holes of the pattern so that the different parts of the design stay separate while you're painting, and distinct once painted. Avoid fussy designs unless you're experienced – you could start with a border and progress to an overall pattern. Suitable paints for stencilling include: signwriters' colours (bulletin colours); Japan colour (US only); artists' acrylics; oil-based eggshell; poster colours; or undercoat. All these can be thinned (or tinted with stainers' or artists' oils) and dry fairly fast. Stains are another option. Experiment on a spare piece of matching wood first to gauge the effect. Spray paint for car bodywork can be used too, but it is difficult to manage.

Apply the stencil to bare boards or a painted ground. You need only a little paint. For transparent, brilliant colours, several thin coats are better than one thick one. Protect stencils with several coats of transparent dressing, make sure you choose one compatible with the paint you have used.

Spattering produces interesting pebble-like patterns of different shapes and sizes.

Spattering

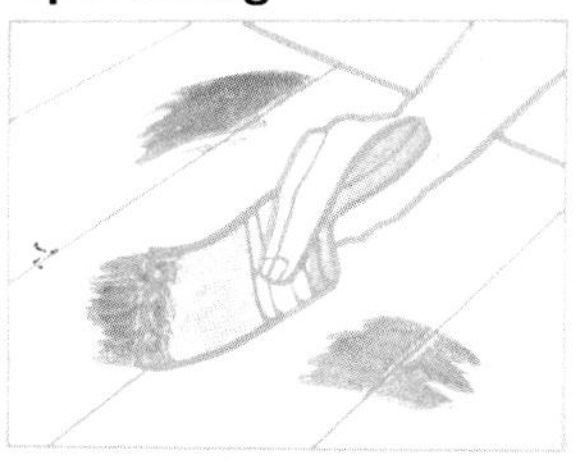

1. Make two colour glazes. With a paintbrush, slap down one colour at a time.

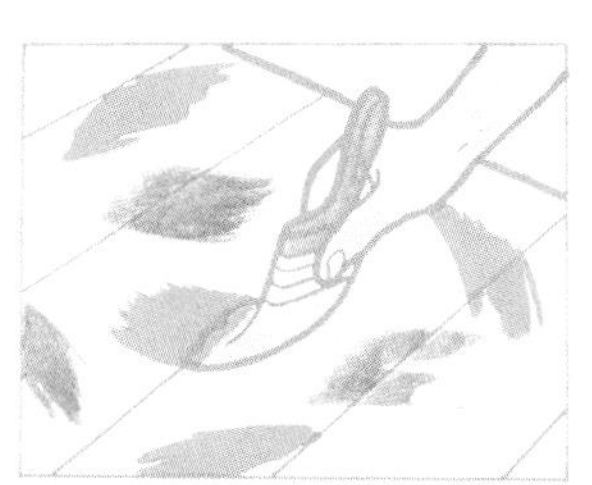

2. With a clean or new brush, slap down the second colour as before, in the spaces left.

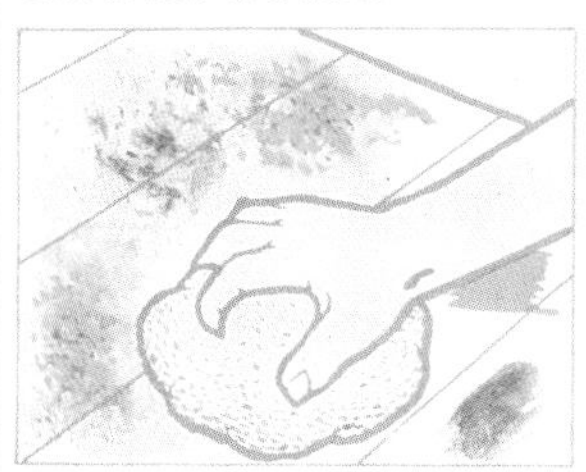

***3.** With a sponge, stipple over the wet colours, scrabbling them together.*

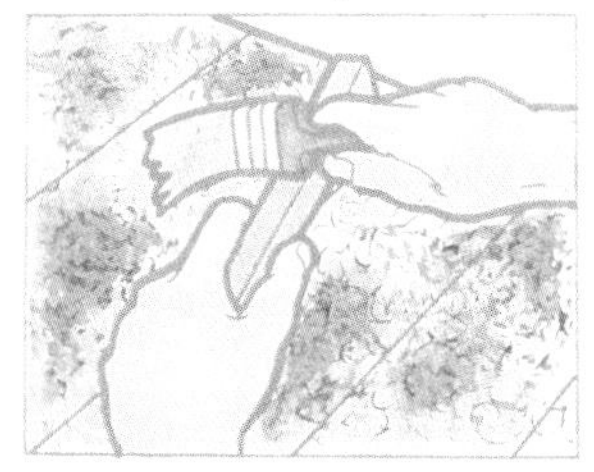

***4.** Knock paintbrush dipped in white spirit on wood so that it spatters on the wet paint.*

Spattering

Paint on a base of standard undercoat, acrylic primer or thin emulsion, followed, when dry, by white or off-white oil-based non-gloss paint. When dry, sand lightly. Rub with a very thin coat of three parts white spirit, one part linseed oil and a drop of liquid driers.

Make two glazes by mixing one part white oil-based paint, two parts white spirit (mineral spirit) and tinting with artists' oils or universal stain. Work small square areas at a time. Slap down one glaze in chequers, then the other in the spaces; scrabble over the still-wet colours with a natural sponge to mix the colours roughly together. Flick on white spirit (mineral spirit) so that small holes appear, and grow, to resemble stone-like patterns. For variety, methylated spirits makes bigger, blurrier shapes; water, neat, little ones. Or, flick on thinned paint or glaze after trying out colour mixes.

Once the floor is quite dry, varnish with several coats of slightly thinned clear gloss polyurethane varnish, sanding lightly after the last but two and the last but one coats.

Marbling

A 'marbled' look can be achieved by painting the floor in the same base as for spattering, then using artists' oil colours in toning shades. Mix two parts of the same base paint with one

Marbling simulates the cold, richly veined quality of natural stone or marble.

Use combing for a textured finish or, with a second colour, for contrast.

Marbling

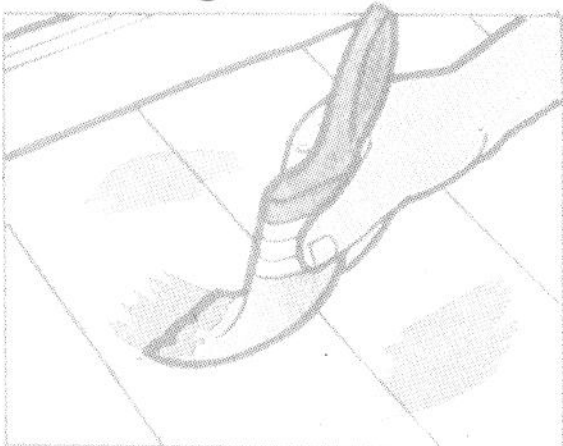

1. Apply paint in the same way as in spattering, using a 50mm (2 in) decorator's brush.

2. Using a natural sponge, 'pounce' the wet paint to stipple up the surface.

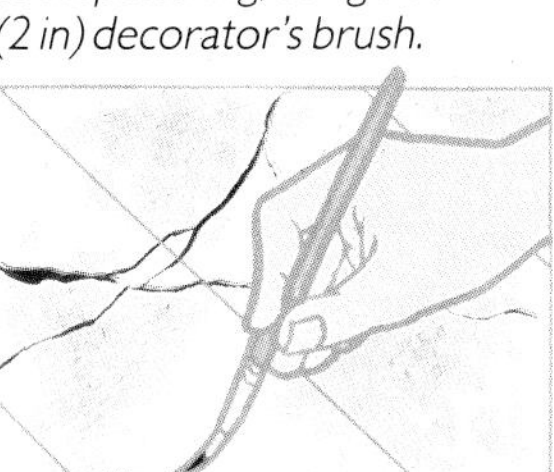

3. Mix a second colour with a touch of black, and paint in veins with a pointed brush.

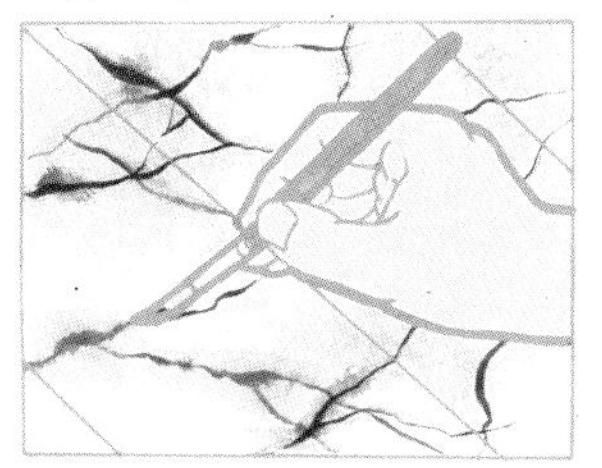

4. Soften and blur the veins with a dry brush, imitating the appearance of real marble.

part white spirit, add a squeeze of one oil colour plus a little black. Mix, apply with a 5cm (2in) decorator's brush then 'pounce' using a natural sponge, to stipple up the surface. Mix a second chosen colour with a tiny touch of black and paint in a wandering vein with a small pointed artist's brush. Soften and blur the vein with a dry brush, but keep it vein-like. Alternating with another shade, paint in smaller veins: the lines should wander in and out of each other to imitate the lines in real marble. Continue sponging and dry brushing gently, but don't overdo. Varnish with polyurethane twenty-four hours later.

Combing

Paint the floor a good opaque colour – you can either comb the last coat for texture, or comb a different colour for contrast. Comb in stripes or squares (which need be only roughly accurate), going in the same direction or in different directions to make a random or regular pattern. Use a special graining comb, a comb used for spreading tile adhesive or a comb homemade from any rigid material. The wider the comb, the quicker you'll work. Lay on combing colour, comb with straight strokes or curves, wiping the comb clean of any excess paint as you go. When the paint is hard and dry, coat with polyurethane.

NEW WOOD

Right: An immaculately laid zigzag pattern of new, pale hardwood complements the classic setting and modern furniture. The effect is decorative but not intrusive.

New wooden floors are classic, contemporary, attractive, warm underfoot and durable. They can be laid structurally in soft or hardwood strips or boards of varying thickness, or in factory-made 'tile' systems. Only stiletto heels and cigarette burns are likely to cause lasting damage.

Selecting flooring timber

If you plan to go to the trouble of laying a new floor, you must also go to the trouble of selecting good quality, properly dried and seasoned timber. The harder softwoods and a wide range of hardwoods are all suitable for floors. Avoid softwood not impregnated against rot and woodworm; softwood with too high a resin content that will 'weep'; sapwood and the heart of hardwood logs (too soft and likely to split, shrink or twist); badly dried or badly stacked timber; over-warped or cracked timber; timber with dead knots; damp timber. Kiln-dried timber should be ready to use. Old timber, occasionally available when buildings are demolished (and sometimes at bargain prices), may be well seasoned and solid – or it may re-saw badly.

Flooring wood must contain roughly the same amount of moisture as exists in the atmosphere of the room in which it will be laid. Wood that is too dry may regain moisture from the atmosphere during the summer and swell; wood that is too wet may dry out *in situ* and shrink. Where there is underfloor heating, the moisture content must be in the range of 6 to 8 per cent; if there is ordinary central heating, the moisture content should be between 10 and 14 per cent. Wood for flooring also must have a density (at 15 per cent moisture content) of around 700 or more – this is measured on an internationally agreed scale.

Once you have selected your timber, check in which widths it is available. Wide boards look rugged, reminiscent of ships' timbers. Narrower boards will give a more refined look. Most wood comes in standard widths, so specifying odd sizes may mean a high degree of wastage.

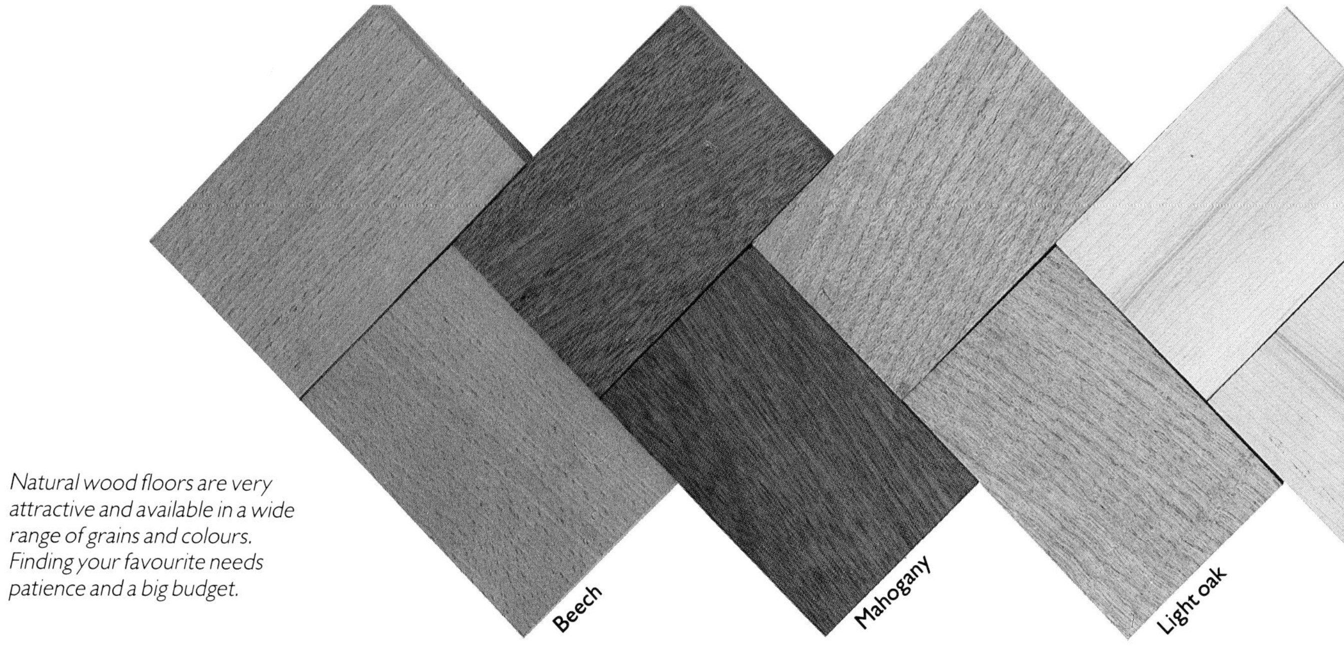

Natural wood floors are very attractive and available in a wide range of grains and colours. Finding your favourite needs patience and a big budget.

Walnut
Maple
Elm
Dark oak
Pine

Ready-made wood floors

Manufactured wood floors include hardwood strip, wood mosaic (usually in a basket-weave pattern) and wood block. Price varies according to quality and finish – but compares favourably with quarry or ceramic tiles. Many systems interlock and come in different stains or natural colours so that patterns can be made. They are often ready sealed by the manufacturer.

Some types are best professionally laid and finished; others are easy to fit yourself. All of these floors must be laid on completely smooth, level, damp-free sub-floors – either concrete, or timber, ply, chipboard or hardboard. For ground-level floors with a concrete base you will need a damp-proof membrane, overlaid with a self-smoothing screed. The screed material is poured on to a clean, grease-free floor, levelled by trowel then left to level itself by settling. Do not lay this type of floor if you haven't protected against damp. Once laid and sealed it is easy to maintain by sweeping, wiping and occasionally polishing.

Hardwood strip

These floors are composed of strips of solid hardwood or veneered softwood. Provided the veneer is thick enough this flooring should be perfectly durable. If there is underfloor heating, a stable wood is best – merbau, iroko, or teak. Other good woods are oak, hard maple, keruing and mahogany. All should be treated with preservative.

Fixing will vary from system to system: on a sub-floor adhesive is often used. Do not push strip against strip if laying with adhesive: the adhesive will seep up through the joints. Instead set the strip down carefully to butt the next one, and wipe off any excess adhesive immediately.

Other systems interlock and can be dry-laid except for the last few panels. Some can be fixed with panel pins if there is a wooden sub-floor: drive pins diagonally through the tongue of each strip or through the flaps which protrude from the sides. Beware, however, of knocking a long pin through a pipe or wire beneath the boards. Strips can also be secret-nailed to softwood joists or battens running at right-angles to the strip flooring.

Tongue & groove fixing

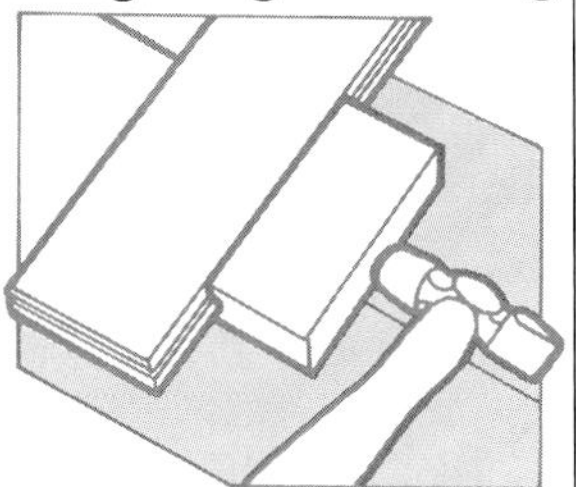

Strips can be gently knocked into place using a spare piece of tongue-and-groove flooring as a buffer.

To attach strips to a timber sub-floor, 'secret nail' at a 45° angle through tongue.

Laying wood strips

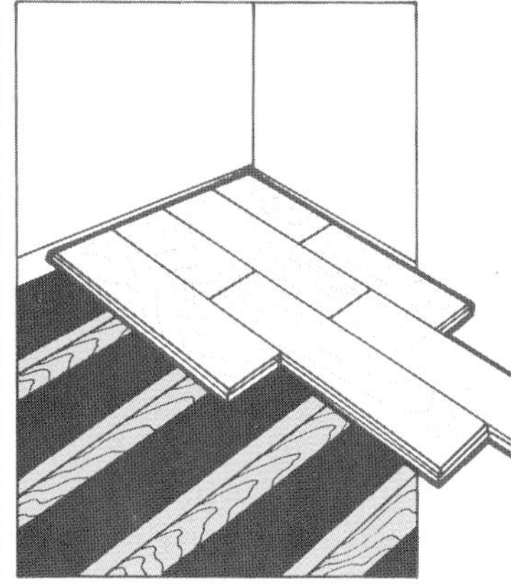

1. *Lay the wood from the corner at right angles to the joists.*

Laying wood blocks

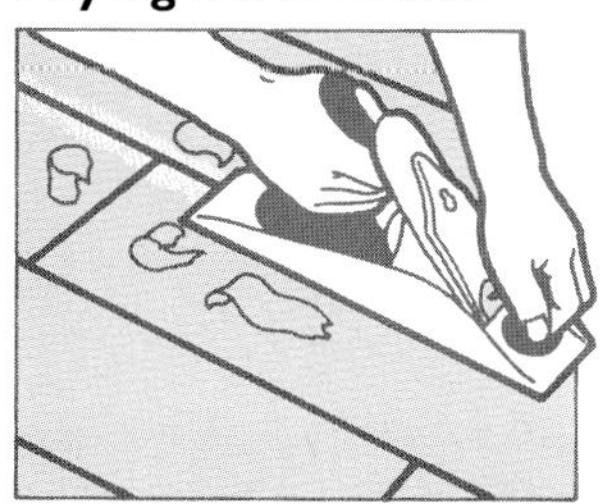

1. *Ensure that the surface is flat. Any raised edges should be levelled with a plane.*

Solid wood blocks. Very hard-wearing. Particularly suitable for industrial use.

2. *When ends are tongued and grooved, they can meet in the spaces between the joists.*

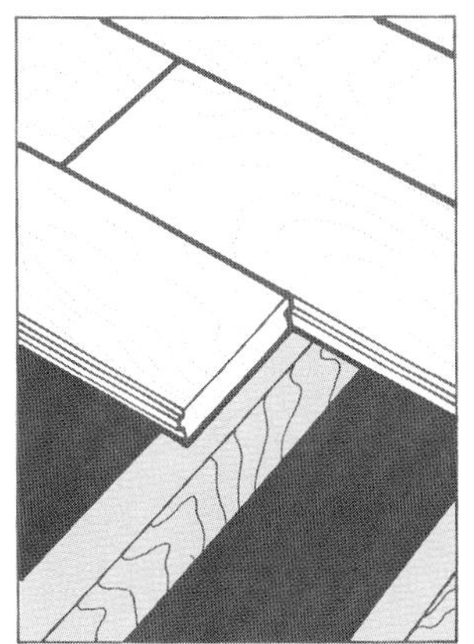

3. *Flat ended strips must join on a joist. Strips are usually fixed by secret nailing.*

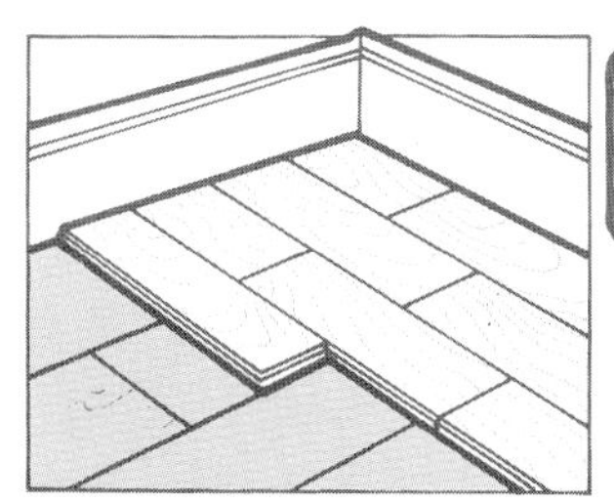

2. *Blocks should be laid at right angles across existing boards. Joints should be staggered.*

3. *A bitumen adhesive can be used instead of nails. It may cut down squeaks.*

Laying mosaic tile panels

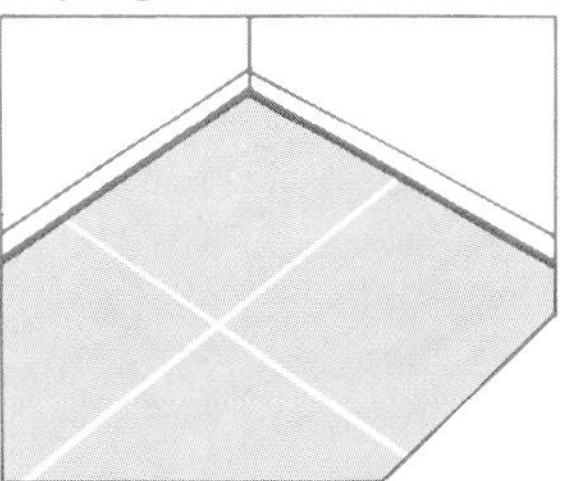

1. *Stretch chalked string from the centres of opposite walls, crossing at right angles. Snap to transfer chalk line to floor.*

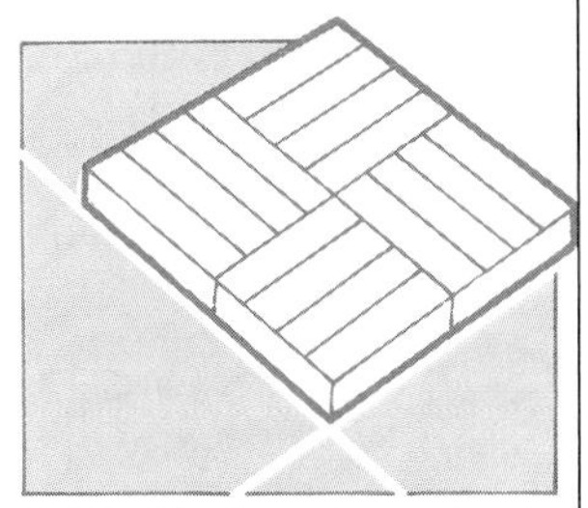

2. *Working from the centre, dry lay tiles to check fit, then attach with adhesive. Leave fractional expansion gap.*

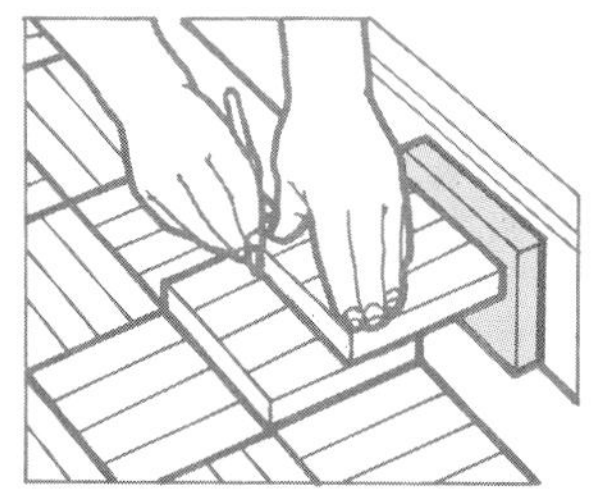

3. *Cut tiles to fit edge. (See page 59.) Keep correct expansion gap using wooden batten.*

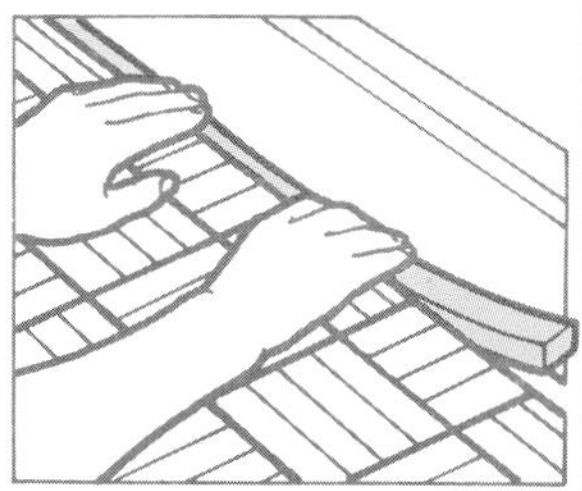

4. *Measure expansion gap. Cut a cork strip to fit it and stick firmly down.*

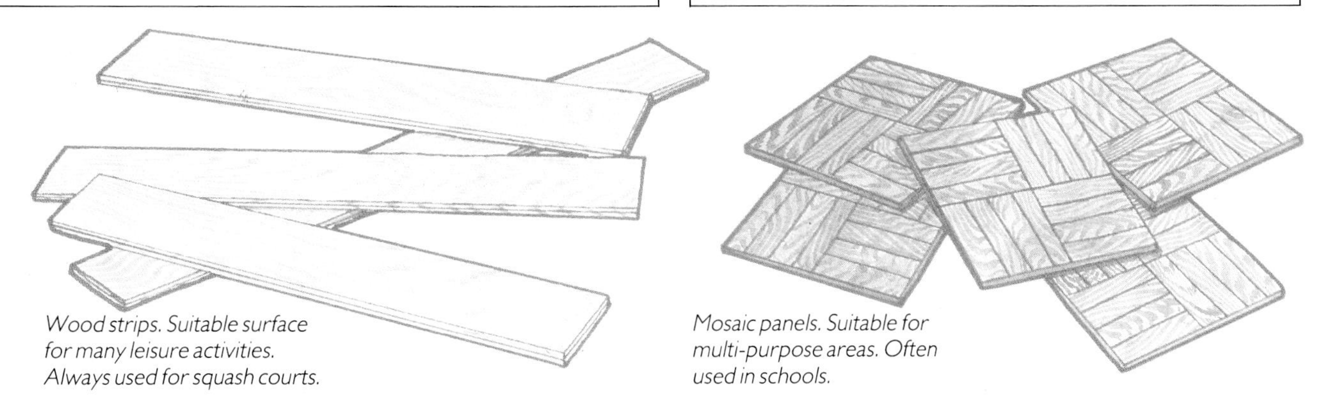

Wood strips. Suitable surface for many leisure activities. Always used for squash courts.

Mosaic panels. Suitable for multi-purpose areas. Often used in schools.

Wood blocks can be laid in a variety of different patterns. Herringbone (left) is one of the most commonly used, but zigzag, basketweave and Dutch (below) are also popular.

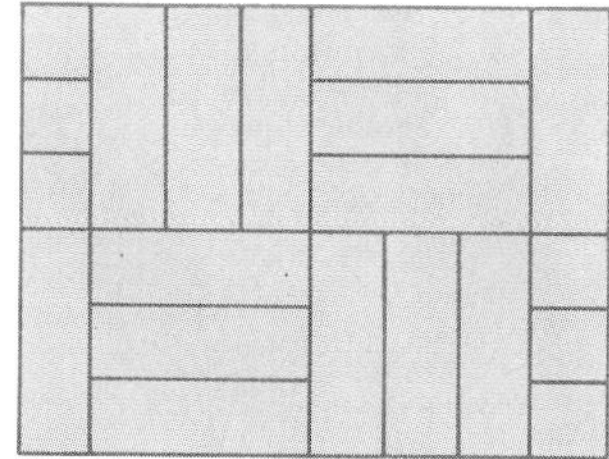

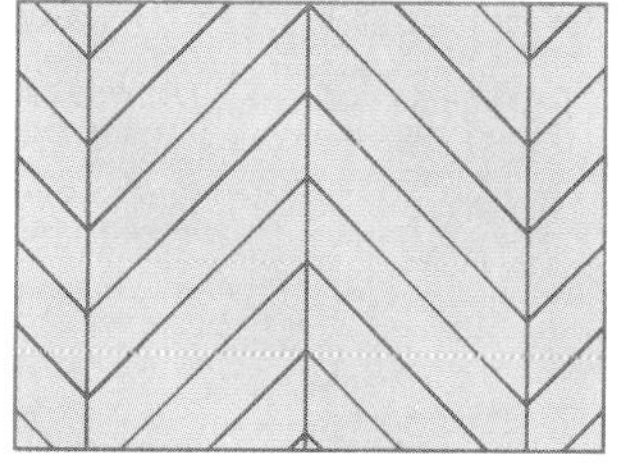

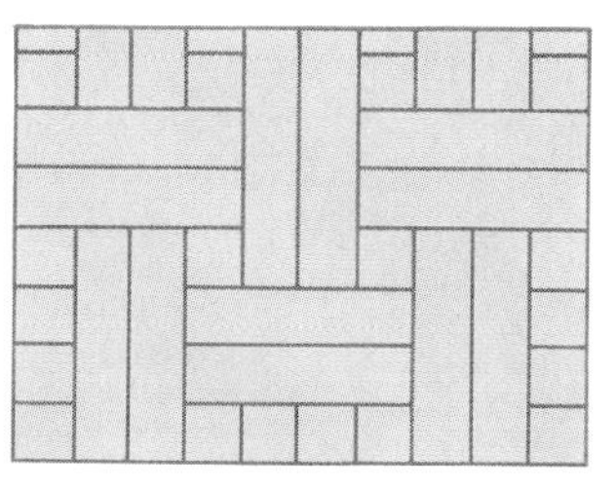

Machine-sand lightly after laying (unless the wood strip is pre-treated), then finish with several layers of wax polish or seal with an oleo-resinous base. Do not use polyurethane on a newly laid timber floor; it needs time to expand and settle down first.

Wood mosaic

Mosaic is made of stable tropical hardwoods such as teak, merbau, sapele and iroko, or in European or American oak, and comes in panels composed of squares, each of which is made of four or five small strips. Panels are often faced in paper to prevent the surface being damaged while the mosaic is being laid; the surface will probably need sanding once the paper has been removed. As with wood strip, avoid using polyurethane until the wood mosaic has had time to expand and settle.

Wood block

Common woods are oak, maple, merbau, iroko, mahogany and muhuhu. The blocks are tongued-and-grooved, and may be laid on a screeded base after being dipped into bitumen rubber emulsion. A wide range of patterns is available.

Wood blocks should be kiln-dried to 6 to 8 per cent moisture content for use over underfloor heating, but the figure may vary with humidity. Have a specialist check that your choice of flooring is suitable.

Man-made alternatives

Humble plywood has been laid in giant sheets, and varnished to a tough, mirror-like gloss. The resulting finish literally reflects every detail of the ultra-stark, modern interior.

Hardboard

A thin, inexpensive board made from softwood pulp, hardboard can be sealed or painted for a temporary, cheap flooring that will later make a good base for carpet or tiles. It is not very durable in the long term, especially if water gets underneath or the seal wears away.

If using as final floor, lay smooth side up. 'Condition' before use by brushing water into reverse side then stacking boards flat, back to back, in room they will be laid in for no longer than forty-eight hours. When fixed, the boards will dry out and tighten to fit thoroughly. Begin in the centre of room, slotting in the border around the edges.

If laying for lining, lay from the edge of room outward rough side up, to provide a good 'key' surface and so pins don't protrude.

Hardboard floors must be sealed at once, or painted. Subsequently sweep and wipe with a damp cloth if necessary.

Chipboard

Thicker than hardboard, flooring grade chipboard comes in tongued-and-grooved forms or in sheets. It's reasonably durable, an efficient sound insulator, warm and resilient, and so is often used instead of solid wooden floorboards in new homes. Avoid using where water may be spilled. Once chipboard becomes wet, it is permanently weakened.

To lay, use woodworking adhesive to fix tongues into grooves, removing any excess adhesive immediately because of the dangers of permanent staining. In fact, it's best to paint on the first coat of sealer before you even start laying to minimize the risk of staining. Provide access traps if you need access to underfloor services. Use nails or screws at least several times the thickness of the board at intervals around the edges. Where board edges do not coincide with joists, support on noggings.

Plywood

Plywood with various timber faces can be cut into squares or bought ready tongued-and-grooved and laid like any other wood – on boarding in older houses but even directly on joists if you want to try it as a new floor. It must be of the right thickness (check with your supplier). With a good finish it can look quite classy, interesting but neutral, or it can be painted or stained with colour, which it tends to take in a rather streaky way unless very carefully applied – although such an effect can be interesting. It won't stand heavy traffic and must be thoroughly sealed or treated as soon as it is laid.

Ask the supplier to cut the sizes and shapes you want. Fix with serrated-head hardboard nails. Lay on boards, joists if sufficiently thick, or bed in bitumastic on screeded concrete.

Plywood can be sealed, or painted and stained first and then sealed. Subsequently sweep and wipe with a damp cloth to keep it clean.

TILES

Tiles represent a convenient way of manufacturing and handling the type of heavy or brittle material that would be difficult to make and manage in sheet form. Well-laid tiles – whether they are ceramic, cork, quarry or vinyl – have a regularity and a rhythm that is classic, soothing and stylish. Available in a huge range of materials, textures, colours, sizes, shapes – and prices – tiles not only make good utility floors for kitchens, bathrooms and laundries but can be used imaginatively in many other areas of the home, too.

Depending on their size and whether they are laid across a room or lengthwise, they can alter the impression of a room's dimensions and emphasize its overall style. Very large tiles give a grand look to living rooms and halls – a crisp white-tiled floor teamed with colourful rugs has a cool elegant, Mediterranean style. Small square tiles, on the other hand, would look fussy in a big room because of the number of intersecting lines but their neat gridded appearance is perfect for a small kitchen or bathroom. Small rectangular tiles have a busy pace that suits high-tech interiors,

There are many ways to vary the look of a tiled floor. Borders in a contrasting colour, size or material lend definition. Patterns, from the simple and formal to more complex arrangements, are particularly effective and can be used to emphasize a particular area of a room or highlight a special feature.

Practically speaking, tiles also come in enough variety to meet most needs. Some are warm underfoot, such as cork and rubber; others hard and cold, notably ceramic and quarry. Some are lightweight, others so heavy that the subfloor must be surveyed to check that it can bear the load. Some are easy to lay; for others you should call in professional help.

Right: Cork tiles polished to a high sheen take on an elegance which belies their hard-wearing quality and modest cost. A natural material, cork is a good choice, both visually and practically.

Far right: White ceramic floor tiles with contrasting trim make a pretty and easily cleaned floor for a kitchen that opens to greenery.

Calculating tiles
To calculate the amount of tiles neded, work in metric if the tiles are made in metric dimensions, or imperial if the tiles are imperial. Measure the square area of the room (see Measuring for Carpets page 32). Work out how many tiles fit into a square metre or yard, and multiply this by the total number of square metres or yards necessary. Allow extra for breakage and wastage. If the room is irregularly shaped, draw a plan of the room on graph paper. Map out the central area, calculate the number of tiles needed, then work out the number of tiles needed for alcoves, bays and odd corners, and add the calculations together for the total number of tiles. If you are using coving tiles around the edge of the floor, calculate these separately.

Laying hard tiles

When laying a hard tile floor, it is particularly important to remember that the floor covering will only be successful if the floor beneath is properly prepared. Ceramic and quarry tiles, which have no flexibility, will be uneven unless laid over a completely level surface. Timber floors should be covered with hardboard first. Concrete sub-floors need a cement or sand screed. Tiles can then be bedded in mortar – the correct mix of which is essential – or with adhesive. Hard tiles are not as easy to lay as soft tiles and are more expensive on the whole, but they are very hard-wearing and should last for a long time. Ceramic tiles are not easy to cut and instructions should be followed closely. It is worth investing in the correct tools. The wrong cutters can very easily cause broken and damaged tiles.

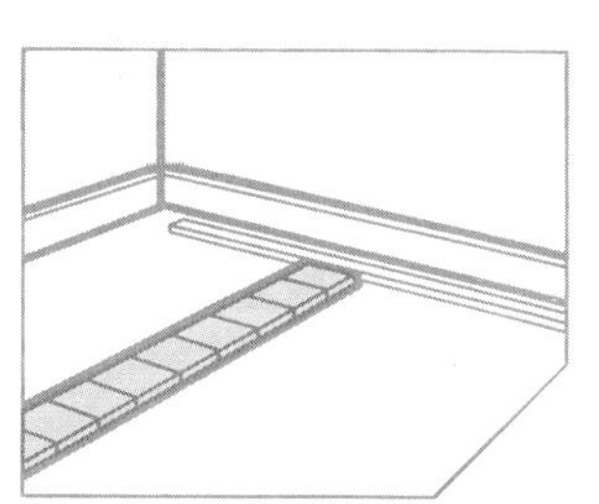

1. Mark line at right angles from door centre to far wall. Dry lay using spacers. Adjust so border tiles not too narrow. Fix batten at 90° where last full tile ends.

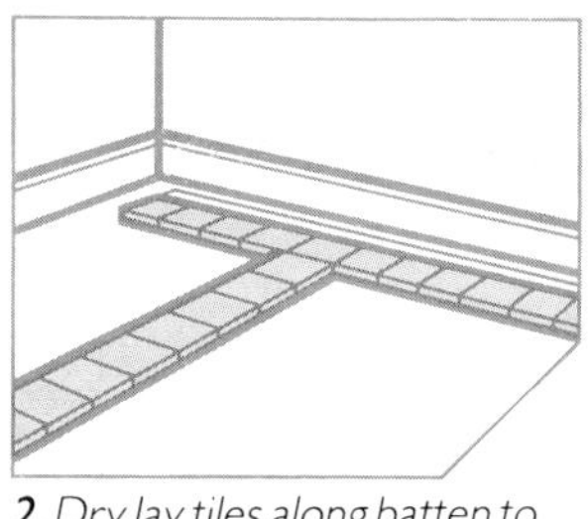

2. Dry lay tiles along batten to either end, again adjusting so border tiles not much less than half-size. Fix second batten at right angles to the first.

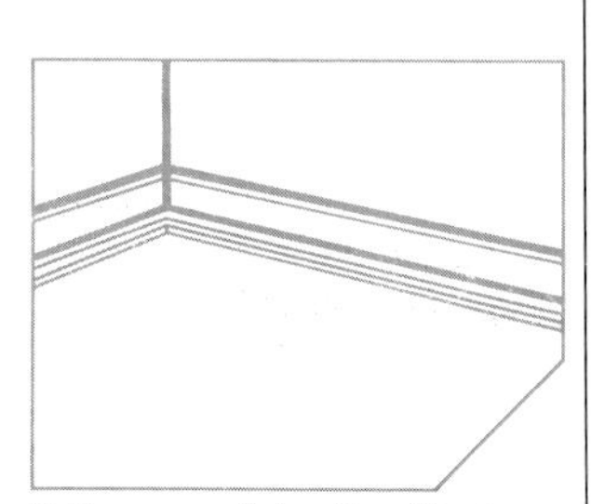

3. Remove all dry tiles. Start tiling in corner formed by the two battens. If using mortar, ensure mix is correct. Otherwise adhesive can be used.

4. With a notched spreader, spread a thin layer of adhesive across a small area. Press down tiles using spacers to separate them.

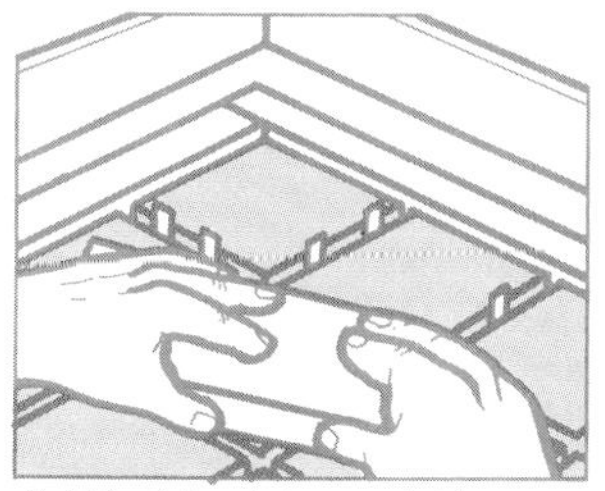

5. Work back towards the door along both battens. Ensure that the surface is quite even by laying a batten across it. Or use a spirit level.

6. Using a set square, periodically check the angle of the tiles. Leave them for 24 hours. Then remove the spacers and finish the borders.

Cutting ceramic tiles

1. Score a firm line through glazed surface and edges of tile, running tile cutter against a rule. Hold unwanted section over table edge. Snap off.

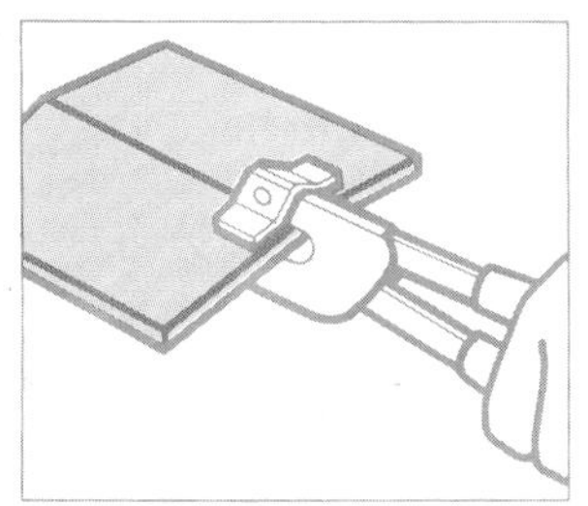

2. Or, when the tile has been scored, use plier cutters. Centre and press evenly on both sides so that the tile snaps in the required place.

Grouting

1. Grouting can be bought ready-made or in powder form. Spread it over tiles with a malleable spreader or sponge until all the joints are well filled.

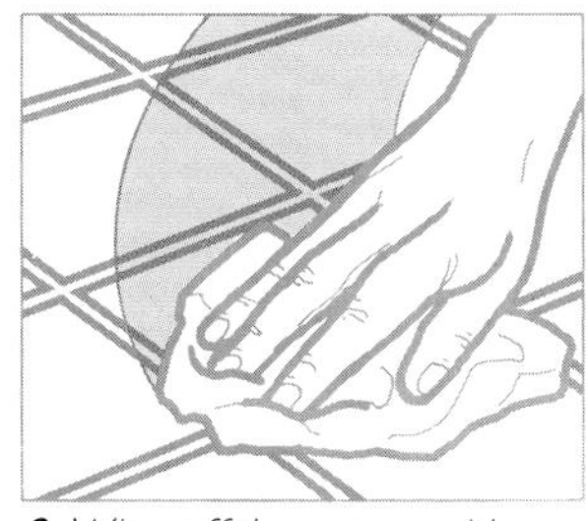

2. Wipe off the excess with a damp cloth before it sets. Finally, remove any hardened grout left on the surface and buff up with a soft, clean cloth.

Laying soft tiles

As a general rule, soft tiles are rather easier to lay than hard tiles. They are lighter to work with and much easier to cut to shape. Like hard tiles they need to be laid on an even surface and a chipboard or hardboard covering provides an ideal base – certainly better than uneven floorboards. Because they are light they are suitable for use in rooms with suspended floors, such as children's bedrooms and bathrooms on first or second floors, where support may be insufficient for harder, heavier tiles. Many cork and vinyl tiles are self-adhesive which makes them easier to lay, although care has to be taken to position them correctly before pressing them down. Soft tiles are now surprisingly durable, and stand up well to everyday wear and tear. Cleaning instructions must be followed as the wrong cleaner can damage the tile.

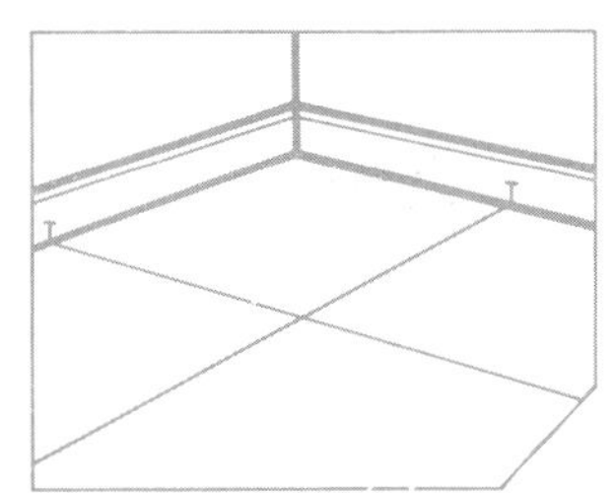

1. Fix chalked line between centres of both pairs of opposite walls. Snap line down so chalk marks cross at right angles. (Check with set square.)

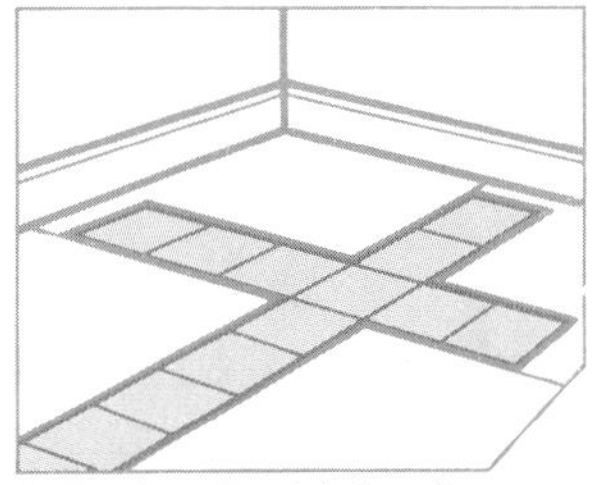

2. Dry lay tiles tightly adjacent along the lines, starting at the centre. Border tiles should be as near full-size as possible: adjust from centre for this.

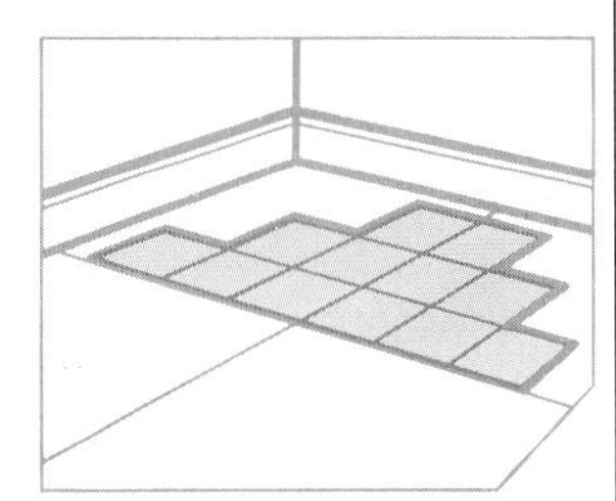

3. Glue the tiles in place, working outwards from the centre in a pyramidal pattern until the field is completed and only the border remains.

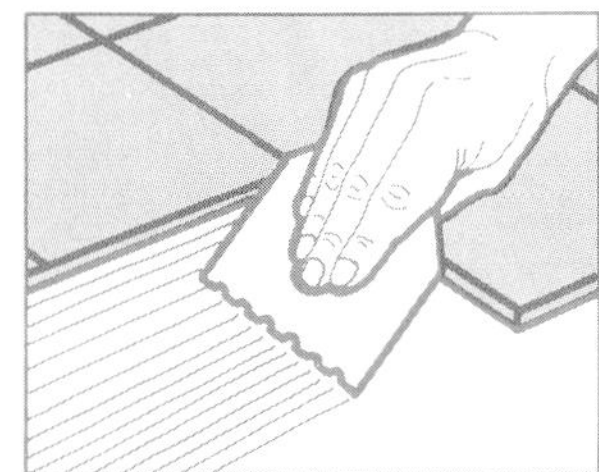

4. Spread glue over slightly larger area than five or six tiles at a time, with notched spreader. Press tiles downwards and tightly together.

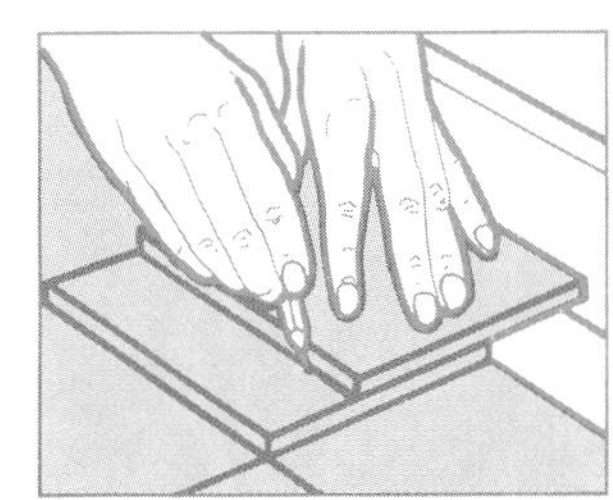

5. For borders, place tile to be cut on the last complete tile, then a third flush with the wall. Using this as an edge, mark the second tile then cut and fit it.

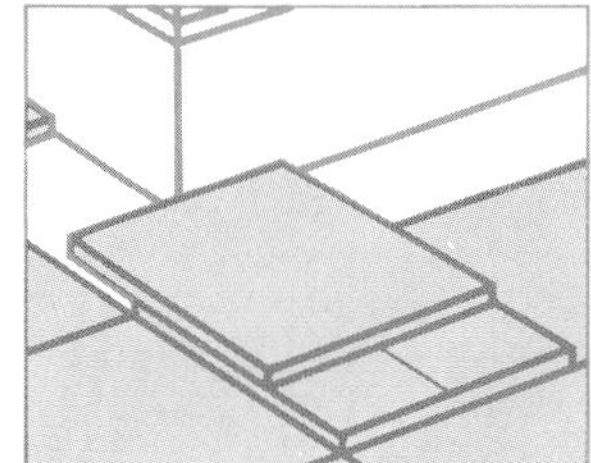

6. Cut corner tiles in the same way. Move the tile first against one wall, then, without turning it, against the other. Mark and cut second tile as before.

Making a template

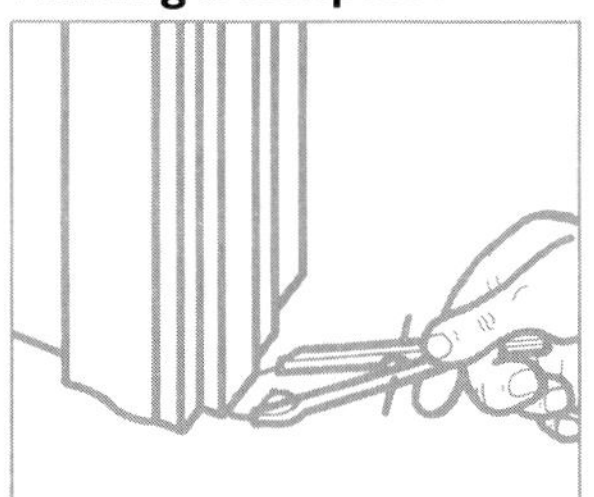

1. Templates are used when cutting around architraves or other awkward shapes in a room. Using a compass, measure shape exactly.

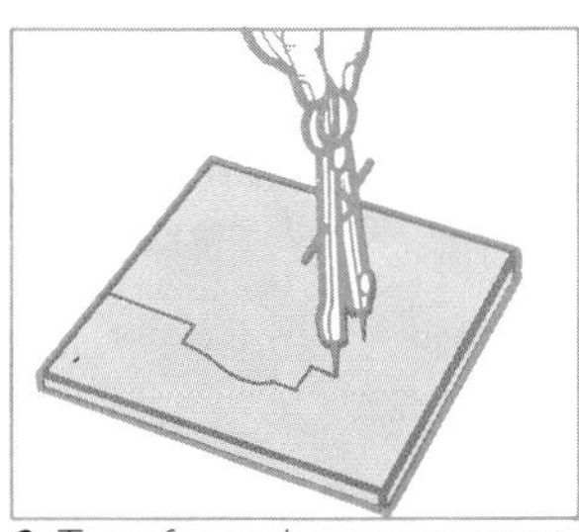

2. Transfer each measurement exactly to a piece of thin card or paper. Cut out the resulting shape carefully. Ensure it fits snugly in position.

3. Place the template on the tile, and mark around it exactly, using sharp pencil or tile cutter wheel. Then, cut the tile with great care.

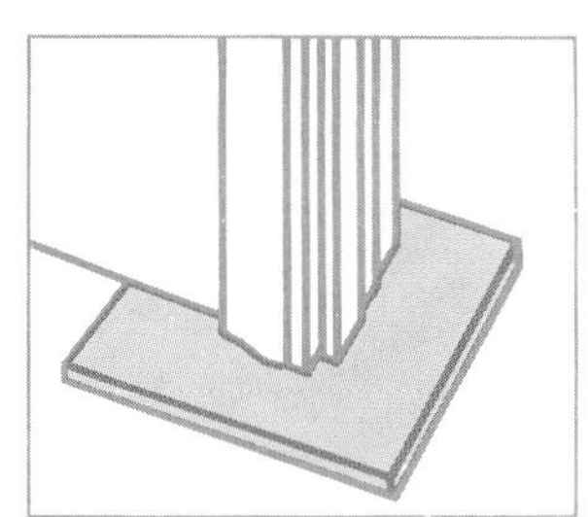

4. Place the tile in position, smoothing any rough edges, and make sure that it fits perfectly. Then stick as with other tiles.

Types of tile

Big, pale quarry tiles make a warm-toned, hard-wearing kitchen floor. Tile variation and uneven grouting result in a hand-crafted look.

Ceramic tiles

To make ceramic tiles, clay dust is pressed into moulds under very high pressure, and then baked at high temperatures. This produces tiles of extreme hardness and strength. The range of styles, finishes, colours and textures is vast: shapes include rectangular, square and Provençal; fiinishes can be shiny, dull, transparent, opaque and unglazed; decoration can be handpainted (irregular and charming) or machine-made (cheaper but precise). Textures range from smooth, embossed and relief patterns to 'anti-slip' surfaces such as raised squares, pinheads or ridges. Interesting colours include the very pale shades and greens, black and blue, as well as the more common warm rustic tones of brown, buff, red and yellow. Browse through the stock of the major manufacturers. Most good ranges have matching coving tiles for a smooth wall-to-floor transition.

As well as being extremely hard, ceramic tiles are also cold and noisy. You will break what you drop, and may well chip the tiles too, particularly if you drop something hard and heavy. Some people find these tiles tiring to stand on. Tiles for outside use must be fully vitrified to be frost-proof.

Since these tiles are heavy and can place a strain on suspended timber sub-floors, it is essential to call in a surveyor to check that there is sufficient load-bearing capacity. Obviously, a small room such as a bathroom will have less of a problem than a larger area. The floor must be level and even. Timber floors should be covered with hardboard first; concrete with a cement or sand screed is also recommended as a sub-floor, but any smooth floor with a latex screed will do.

Ceramic tiles are embedded in mortar (normally a professional job) or fixed with a special adhesive. Use waterproof adhesive in bathrooms, kitchens or utility rooms. Check that tiles chosen for the kitchen are also fat- and oil-resistant.

For after-care, a mild detergent solution will usually do – rinse well afterwards. If the surface of glazed tiles becomes dulled, try a water-softener. Commercial tile bleach will probably solve stubborn stains and refresh dirty grout.

Symmetrical informality is the theme of this entrance hall, achieved by the use of bright blue and yellow ceramic tiles. This bold combination gives impact to the hallway.

Quarry tiles

Quarry tiles, terracotta and similar rustic-looking varieties from countries such as Mexico and Portugal, are made from urefined high silica alumina clay, extruded into a tile mould. They usually come in earth colours such as brickish reds, browns and gold. Though extremely durable, quarries have a very slightly softer surface than ceramic tiles; after years of use they can become a little worn and pitted – but this is not unattractive. There is a wide price range; much depends on thickness. Like ceramic tiles, quarries are cold underfoot and noisy. You will break what you drop.

If the manufacturer allows sealing, use a mixture of one part linseed oil to four parts turpentine. Paint on the seal, let it soak in, cover with brown paper and leave for forty-eight hours. Don't seal with simple polyurethane: it will probably peel away or chip away. Sweep and wash, polish if desired, but take care not to make floor slippery. If white patches appear, wash with a dash of vinegar in water; do not rinse. Clean stains with dilute liquid abrasive cleaner.

Cork

Cork is very versatile and provides a warm neutral base, ideal for displaying bright rugs. If well laid and properly sealed, it is a practical floor for kitchens, bathrooms, children's rooms and cloakrooms, and can also be smart in entrance halls or sitting areas. The colour range is narrow – light honey to dark brown – and tiles tend to be plain, but some are patterned with stripes or squares in a contrasting colour.

Made from natural cork, compressed with binders and baked, these tiles come in a variety of thicknesses and densities. It is essential to use flooring grade. Good quality cork looks dense and even, and is springy underfoot but, if you're laying on hardboard and feeling cost-conscious, thinner grades will be adequate. The toughest type of all has a clear vinyl surface. Cork is fairly warm and resilient but not impervious to cuts, cigarette burns or strong chemicals. It may fade after prolonged exposure to strong sunlight.

Simple, quick and light to cut and work with, cork should be laid on a smooth floor, such as hardboard, rather than uneven floorboards. Store tiles flat and dry in room in which they are to be laid, for a day or two if possible.

Proper sealing with several coats of polyurethane is essential. Wear a mask if the sealer is a strong one. A light sanding before the last coat is a good idea. Allow the seal to 'cure' thoroughly for a day or so after the seal is dry. Even pre-sealed tiles will benefit from a coat or two of seal to prevent liquid seeping between the joints, however tightly you have butted them together. Sweep and wash lightly; occasionally emulsion polish. Avoid spirit-based chemical cleaners.

Linoleum

Once a thin, brittle material prone to cracking and often luridly patterned, good flooring lino today is thick, glossy and durable. Made from linseed oil, ground cork, wood-flour and resins, baked slowly at high temperatures and pressed on to a jute, hessian or fibreglass backing, it can be plain, mottled, marbled, patterned or glittery. Many architects and designers value its toughness and use it to make 'inlaid' patterns with different plain colours – sheet lino is better for curved shapes but tiles could be used to make interesting geometric ones. Fairly warm underfoot and resilient, lino also has a good resistance to dilute acids, alkalis and household chemicals but will eventually rot if water gets underneath. Since it is slightly thicker than vinyl, sub-floor deficiencies will be less damaging, and it can be used over underfloor heating.

Lay on a smooth, dry, level sub-floor – on hardboard or chipboard rather than on floorboards, which may vary in position and cause the lino to crack along ridges. Butt tiles close up against each other – lino seams get tighter with age, which prevents liquids from penetrating.

No sealing is necessary but lino can be polished. You can use the manufacturer's own dressing or a special lino dressing. Sweep and polish lightly. Dilute washing soda can be used occasionally to bring up surface colour. Rub away crayon marks with silver metal polish; try emulsion floor polish on scuff marks.

Plastic

Plastic flooring, borrowed for high-tech homes from sports halls and swimmming pools, is hardly classic but can be a lot of fun. It tends to come in plain, strong colours – black, white and grey, and bright primaries – and is available in sheet form and matting as well as open-work grid or duckboard-type tiles, some of which snap together. While not necessarily cheap, it is a surprisingly comfortable surface. Unless ultraviolet light stabilizers are incorporated, it might fade if continually exposed to strong sunlight. Lay on an impervious sub-floor as water will seep through and take a while to evaporate.

No sealing is necessary. Wash with plain hot water and soap-based cleaner if stained. Don't use heavy abrasive cleaners, chemicals, acids, petroleum, undiluted bleach or acetone. Dirt will fall through holed varieties, and is difficult to remove without a powerful vacuum cleaner.

Rubber/synthetic rubber

First used in industrial and contract work where its tough, non-slip, burn-resistant and soundproofing surface made attractive railway station and airport floors, rubber tiles are

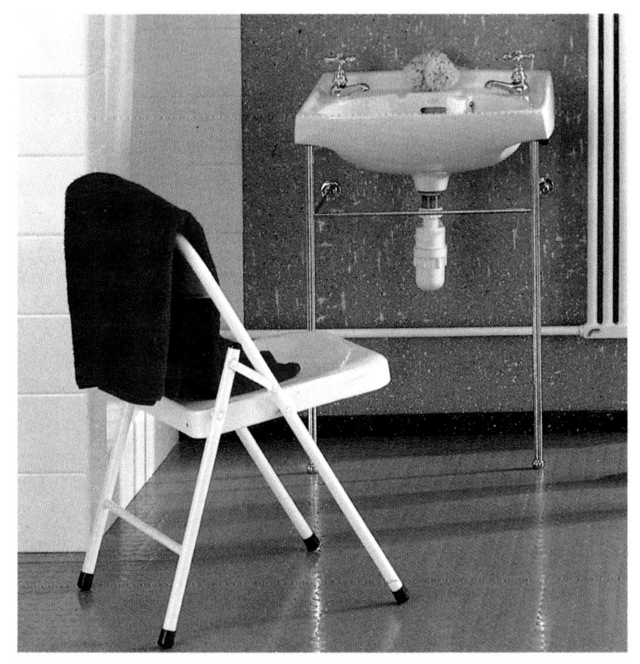

Studded rubber floor tiles make a smart, water-resistant non-slip floor. They are ideal in bathrooms but less suitable in kitchens as sticky spills and crumbs are hard to remove.

increasingly popular in the home, especially for utility areas. Made of different blends of natural and synthetic materials, rubber varies in grade, thickness and hence price, but is rarely cheap, although some thinner, easy to lay, but slightly stretchy varieties cost no more than a medium-quality vinyl. These tiles are normally plain coloured but there are two-toned and marble-effect types. Pattern is generally lent by the texture – round raised studs are common. But beware: when installed in kitchens, there is a tendency for scraps of food to build up tenacious deposits around studs and other relief textures. Narrow tiles are available for stairs.

Thermoplastic

Sometimes referred to as asphalt tiles, these thin, hard tiles are made from asphaltic binders, asbestos fibres, mineral fillers and pigments. A workhorse surface, durable and cheap, these tiles tend to come only in dark browns and blacks, sometimes with a little white vinyl resin thrown in.

Vinylized thermoplastic, with PVC added, is more flexible

Lino tiles give a tough and elegant floor for a classically furnished study. A dull view is counteracted by the brilliant hue and illusion of depth in the pattern of the tiles.

Vinyl tiles come in a huge range of patterns and colours, many imitating other surfaces such as marble or quarry tiles. These are classically simple and subdued in colour.

and highly durable, is available in many more colours and sizes, and slightly more expensive.

Vinyl asbestos

Made from plasticized vinyl resins, asbestos fibres, mineral fillers and pigments, these tiles are tough, flexible, and impervious. They come in a wide range of plain colours, marbled or embossed effects.

Vinyl

Vinyl is an excellent all-purpose material for kitchens, kitchen-diners, bathrooms and corridors but it is not always cheap – the best quality is as expensive as very good quality carpet. 'Vinyl' is a shortened form of polyvinyl chloride (PVC) but the amount of PVC in the material can vary from 25 to 85 per cent – you can usually tell by the price (PVC is expensive). Very versatile, vinyl is produced in varying degrees of flexibility from soft and rubbery to hard, and in tile or sheet form. It can be absolutely smooth but more usually is textured. There is a vast range of patterns and colours, many of which simulate 'real' floors such as ceramic and quarry tiles, sometimes very effectively, sometimes not. Newer vinyls have borrowed the idea from industrial flooring of suspending glitter or tiny crystals in the clear surface layer.

Some vinyls have inbuilt underlays or a 'cushioned' sandwich filling layer for greater sound and heat insulation and greater resilience. Waterproof, resistant to oil and fat and most domestic chemicals, vinyl is not immune to burns or to abrasion by grit.

In winter, vinyl must be warmed before laying – cold makes it brittle. To clean, check manufacturers' recommendations – washing with warm, soapy water and rinsing to avoid a dulling film is usual. Try a pencil rubber or emulsion floor polish to remove scuff marks. Do not use petrol, paraffin, white spirit or wax polish. Manufacturers of vinyl always say they wish their customers would not use harsh, abrasive detergents: one of the main causes of surface damage.

Tiles can be used in many ways to vary the appearance of a floor. Intricately patterned tiles can be used in a pre-determined pattern plan. Imaginative decorators can plan their own designs using plain and patterned tiles and experimentally dry-laying them until the desired effect is achieved.

Far left: Clever use of tiles makes an open plan 'corridor' where none exists – the yellow and grey alternative check pattern is switched with black for three rows to emphasize the traffic area. A black border holds it all together.

Left: Here, different uses of the same, small black and white tiles delineate different living areas. Making accurate graph paper patterns beforehand is essential to planning.

This simple, traditional border pattern is easy to achieve. The corner tiles – the same as the other tiles – have been cut, and the pattern rematched.

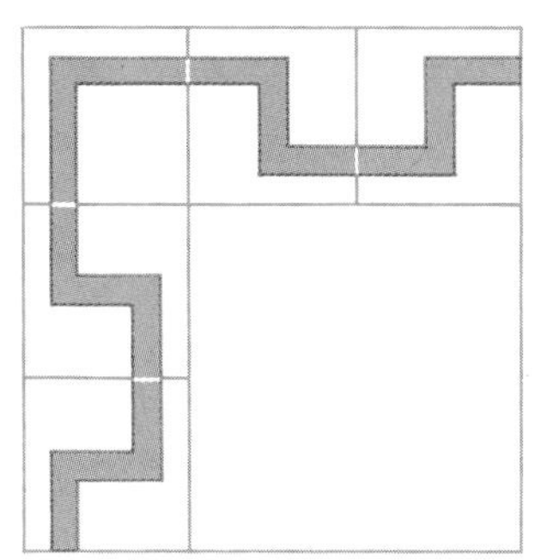

These tiles – a variation on a classic key pattern – could be laid in a number of ways. The corner tiles may be specially bought to continue the pattern.

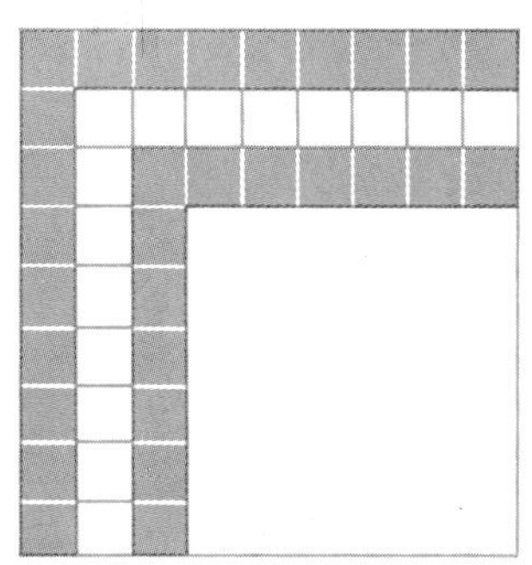

If a room is not rectangular, a geometrical design may look unsatisfactory. In the right setting, this simple border may be stunningly effective.

SHEET FLOORING

The virtue of sheet flooring is that a larger area can be covered quickly and with a minimum of seams – but since the material can be heavy and unwieldy, it is more difficult for the amateur to lay. Vinyl, lino and rubber are all available in rolls as wide as 4m (13ft), which helps to minimize the chances of water seeping between joints and promotes a smooth overall look. Paradoxically, quite a few sheet vinyls are printed with a simulated tile pattern but the aim is just to achieve the tiled look more easily.

Sheet lino is also the best choice if you want to create a floor pattern involving curves since different shapes can be cut out of it. However, dealing with hefty rolls is not easy. Unless you're confident, it is probably a good idea to call in professional help. Nothing is worse than a length of vinyl that has been cut too small or with an embarrassingly wobbly edge.

Before laying, decide which way the sheet will run. Avoid ending up with a seam in the middle of a doorway. If the flooring is laid at right-angles to a window the seams will not be as noticeable. If the sheet is going directly on to floorboards, it should run at right-angles to them, but it is best to lay it on hardboard, to avoid uneven wear. Measure the room to establish how much sheet you need and how wide it should be in order to minimize cost and wastage and match up patterns.

Below: A cool, marbled sheet lino loses its period connotations by being set in a modern high-tech style. It becomes a perfect partner for a chrome-edged chair and picture.

Right: Sheet vinyl in plain geometric patterns makes a smart, easily cleaned kitchen floor. Vinyl is also relatively warm and soft, added assets if there are children in the family.

Laying sheet flooring

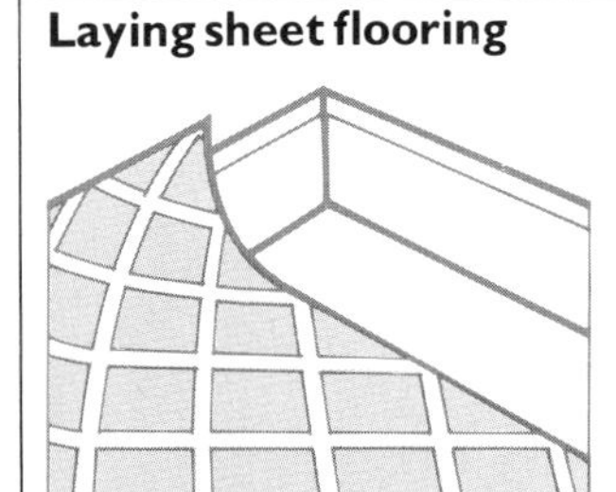

1. Lay sheet along floor parallel to side wall. Leave small gap. Overlap the wall at the end.

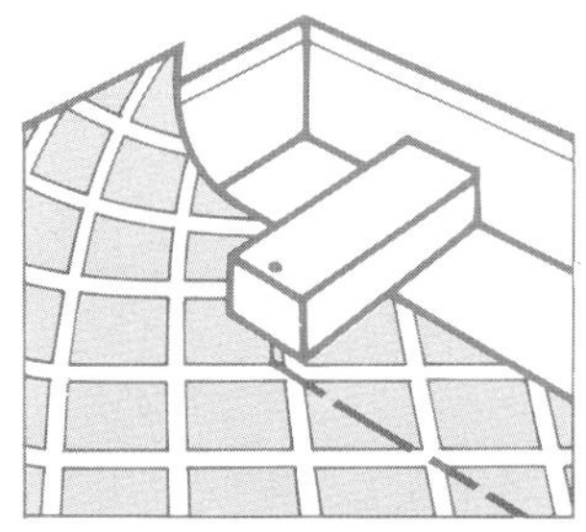

2. Using scriber – wood block with nail protruding – pull along wall scratching sheet. Cut.

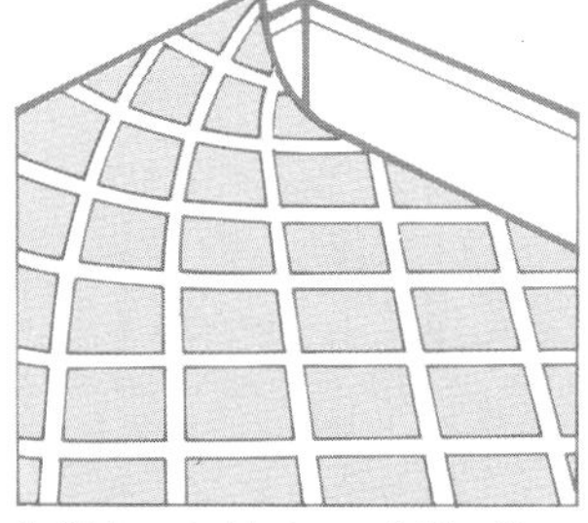

3. Slide cut side to wall. The fit should be exact. Keep the overlap on the end wall.

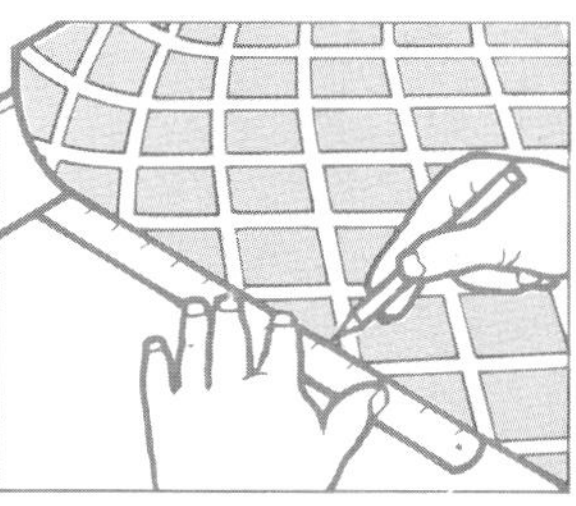

4. To trim ends, measure back a distance of about 200mm from the wall and mark the flooring.

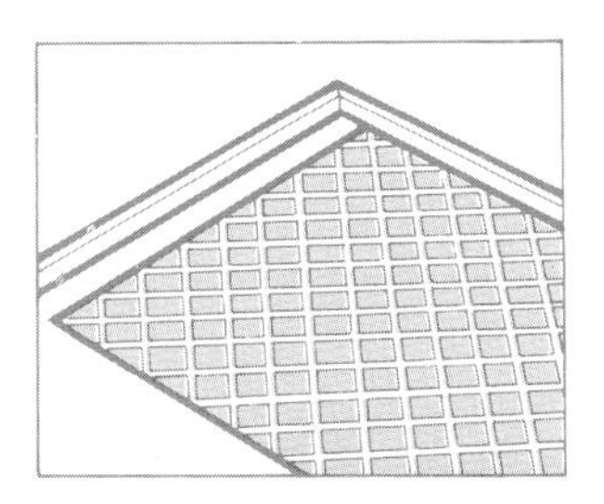

5. Pull the end back keeping it straight against the other wall and lay flat.

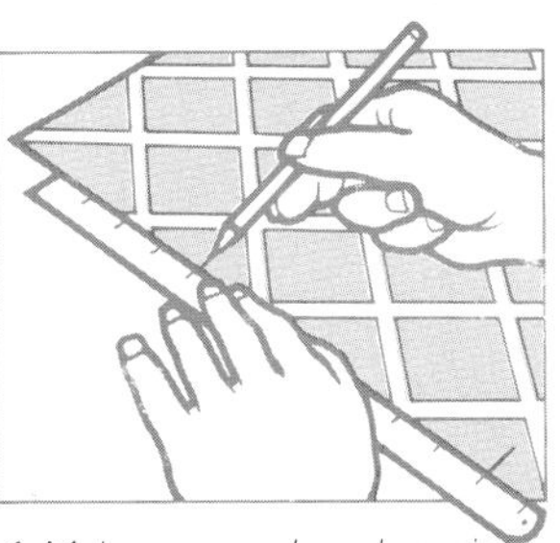

6. Make a second mark, again 200mm, from the first, towards the end of the sheet.

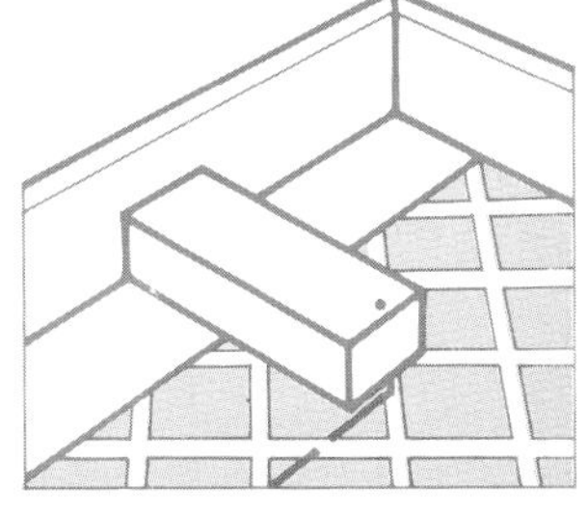

7. Adjust sheet so scriber point falls on second mark. Scratch sheet pushing scriber along wall.

To complete the process, cut along the scratched line and fit the flooring against the wall. Repeat at the other end.

Where it is necessary to lay more than one sheet of flooring, overlap the two sheets slightly, matching the pattern. Leave to settle for a few days, then trim the second sheet. Butt edges. Attach to floor at seams and perimeters with double-sided tape.

HARD FLOORS

Hard flooring ranges from expensive but beautiful natural materials such as slate and marble, to eminently useful ones like brick and the unlovely, but practical, concrete. Durable and weather-resistant, these surfaces naturally find a use outside the home – as patios, terraces, paths and driveways – but they can also be very effective inside, especially for those indoor/outdoor areas where comfort is not a priority – conservatories and entrance halls, for example.

Because of their sheer weight, dealing with any of these materials is not easy, however, and you'll certainly need help. Proper preparation, the right tools, and taking time to make a firm, level base are all important.

If you are laying new concrete make sure you do not breach the damp-proof course in any way (by laying it above the level of the course, for example). If you are laying a replacement ground floor, ensure the foundations are deep enough, of the correct material, that a damp-proof membrane will be incorporated, and that you know where and how drainage, sewers and so on run. To check this you will need professional guidance.

The other essential is correct drainage. Concrete, slabs and bricks must be laid so that water falls away from any structure, either to a nearby 'soft' garden or into a gulley or gutter – gulley channels could make a pattern in themselves. A gradient of 2.5cm in 2m (1in in 6ft 6in) should be enough, but much depends on the specific situation.

Other details to bear in mind: paving should be about 1cm (½in) lower than a lawn to allow for mowing. Any kerb used must be complementary to the paving and the surroundings. It could even be hidden – for example, creosoted wood just below the surface will do the job of holding the paving in place, but invisibly.

A terrazzo floor is beautiful but expensive. It brings classical grace to a dining area, defined by the spot pattern, and overcomes the proximity of the kitchen. A dark border lends definition.

CONCRETE

Cheap and long-lasting, concrete is the ultimate in hard-wearing surfaces. More often than not, you will want to lay it as a base for other types of flooring or in cellars, driveways or areas where appearance is not a prime concern. If it is to be visible, in a garden for example, it must be used thoughtfully or it will live up to its reputation for monolithic ugliness.

There are a number of ways of playing down the natural brutality of the material. In the garden, avoid solid, unbroken areas; small shapes look best. Paths are better with slight curves. Wherever possible mix the concrete with other materials. In a patio, slabs can be alternated with bricks or granite setts, or spaced so that grass, herbs or edging will help to focus attention on flowerbeds, ponds or borders. Colour can also be effective. Bright white concrete could look stunning in a 'high-tech' garden. The Mexican architect Luis Barragan paints concrete slabs in brilliant, soft shades creating a look evocative of the tropics.

Pre-cast concrete slabs are inexpensive compared to stone, and are ideal for an outdoor, urban setting. Here they are laid in a rigid pattern which links with the lines of the quarry tiles in the kitchen, but the visual effect is softened by the presence of plants.

Right: Bright blue, textured paint has been applied to floor and wall surfaces alike, in a bold sweep of colour. It is a brilliant solution to the potential gloom of a shady staircase.

Laying concrete

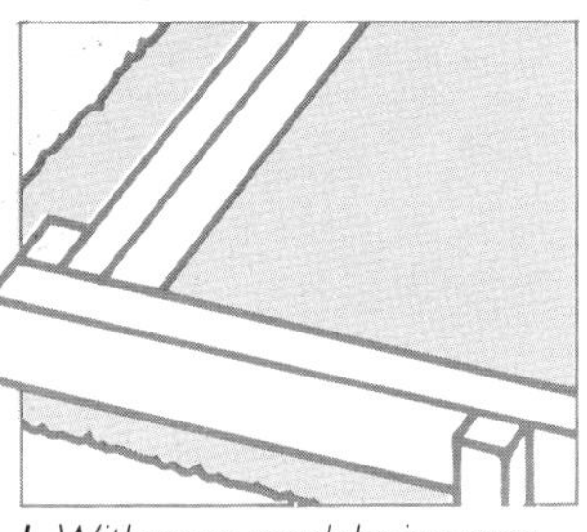

1. With pegs, mark laying area. Sink box of 1in planks. Ensure formwork is level.

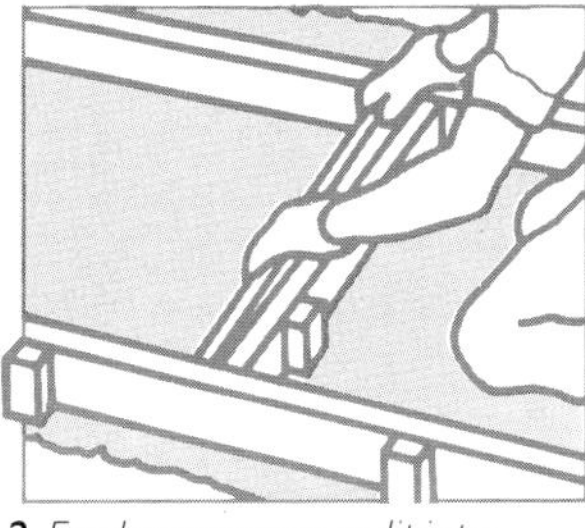

2. For larger areas, split into sections of up to a cubic metre. Insert expansion joints.

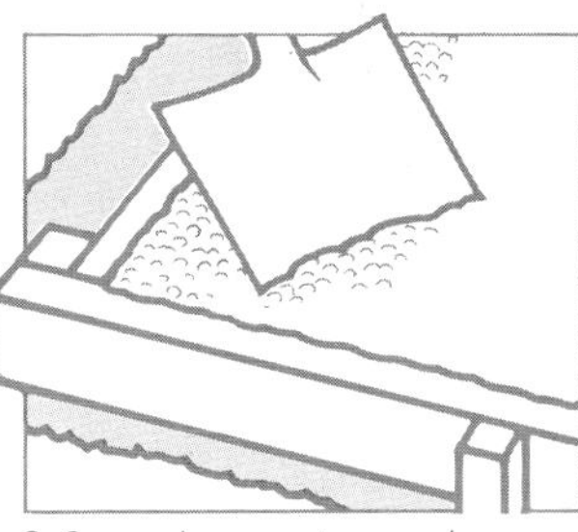

3. Spread concrete evenly. The boards' top will be the concrete surface.

4. Overfill. Rake concrete 15mm (¾in) above the framework to allow for compaction.

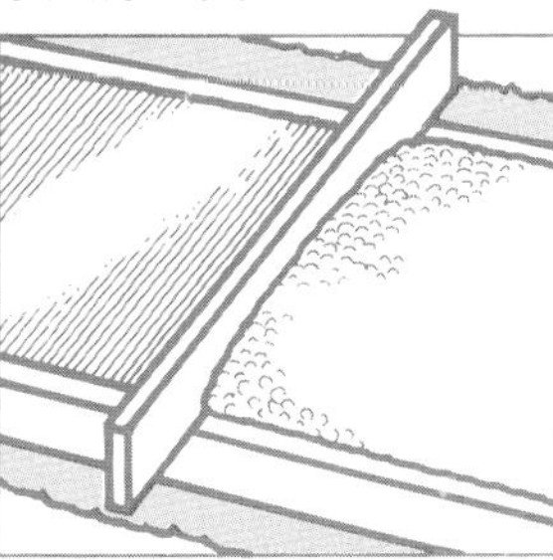

5. Tamp smooth and remove excess, by lifting and dropping edge of plank along surface.

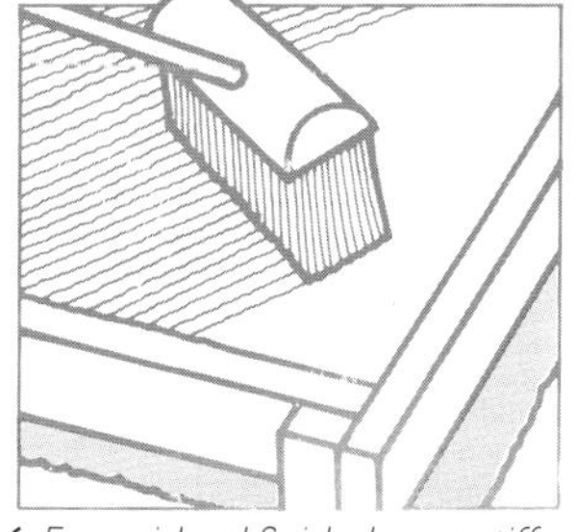

6. For a ridged finish draw a stiff brush over the surface while the concrete is still wet,

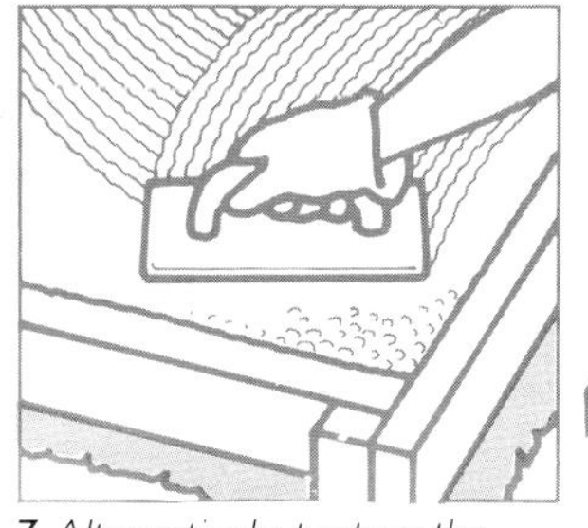

7. Alternatively, texture the concrete with a wooden float, making fan or other patterns.

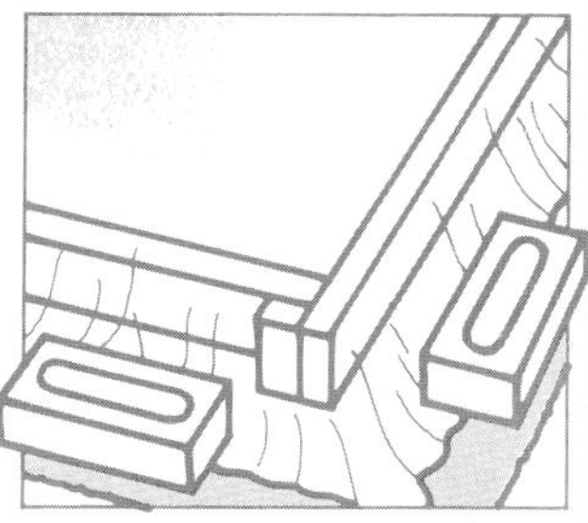

8. Cover concrete with plastic to prevent drying. Secure with bricks. Leave for 7-10 days.

Concrete selection

You can buy concrete in ready-made blocks or slabs of different sizes, shapes, colours and constructions, including interlocking 'bricks'; or you can go to the hard work of mixing your own; or it can be delivered ready-mixed. Concrete is usually a blend of sand and crushed stone or gravel of varying fineness, and Portland cement. The hardening of the cement powder on mixing with water is what binds the mixture together. The crushed components are called aggregate, and are classed by their fineness. After adding water and thoroughly mixing, concrete stays workable for about two hours, less in hot conditions. It takes at least three days to

Concrete composition

CEMENT Ordinary Portland cement or, for brilliant white, White Portland cement.

AGGREGATE Buy fine and coarse aggregates separately and mix, or mixed ballast, but check the quality. 60% of the particles should be over 5mm (¼in) diameter. Avoid ballast that stains your hands when squeezed. Use stone chippings for exposed aggregate finishes.

SAND Sharp, not soft, sand. Dry-packed mixes. Pre-mixed cement, sand and aggregate; expensive but easy to use.

Laying concrete slabs

1. Prepare a firm, level base, then bed each paving slab on five mortar dabs.

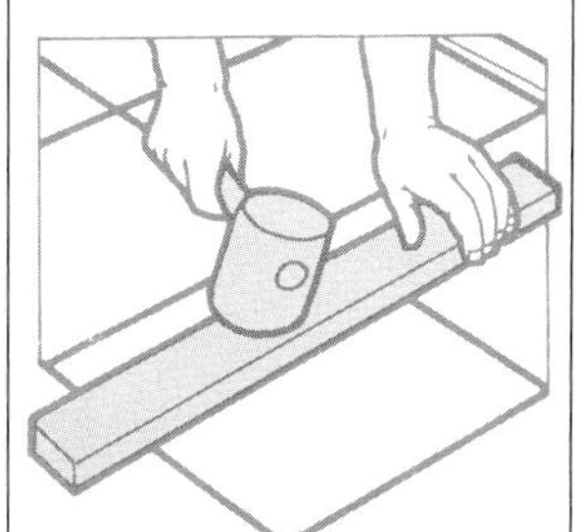

2. With a mallet and batten, knock each slab into place. Check slabs are level.

reach any useful strength, a week or so to reach half its final strength and, for the first few days, it needs to be kept a little moist – it does not harden by drying out.

Preparation

Measure area to be constructed and multiply by thickness to be laid. Work in either imperial or metric units throughout. Add about 10 per cent for wastage. Do not order more cement than you can use in a week, but aggregates will keep. If you find you need more than 3 cu m (3 cu yd), you'll almost certainly want it delivered ready-mixed, or the job done professionally, especially if the concrete can't be delivered right to the site and you have to transfer it by wheelbarrow – 1 cu m (1 cu yd) represents about forty barrow loads, and a lot of pushing.

To prepare an outdoor site, strip grass, roots and topsoil back (roots of trees and plants you want to keep will have to be moved or rerouted carefully). Concrete can be laid directly on good, compacted earth, but will need a layer of gravelly stone (try not to use builders' rubble) if earth is of clay or peat or the site is for a drive or as a foundation for something heavy. Make the site at least 15cm (6in) larger than the proposed size. Dig to required depth and compact soil. The base must be firm and level.

NATURAL STONE

Crazy paving

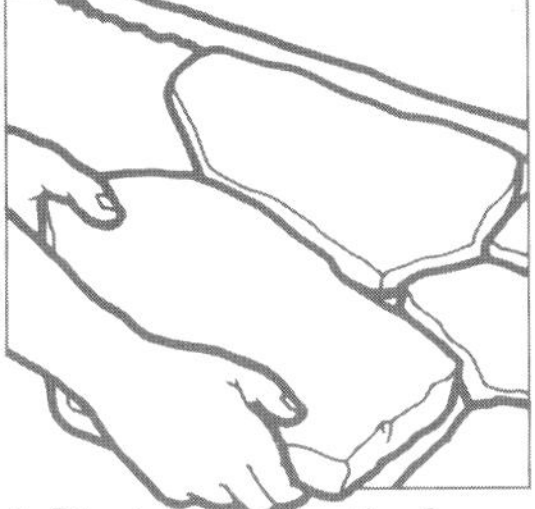

1. Dig down to depth of stones. Flatten. Dry lay. Use regular stones for edges.

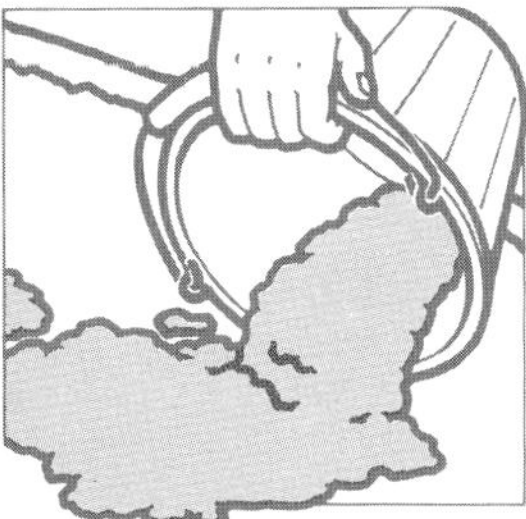

2. Remove a few stones and put by. Lay down 25mm (1in) of mortar. Replace stones.

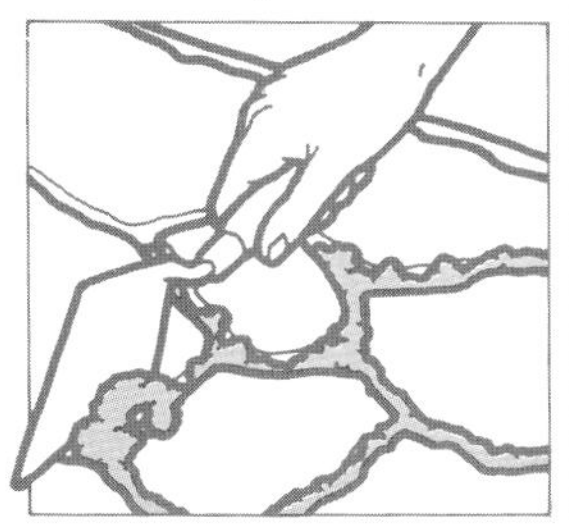

3. When the slabs have been laid, with small gap between each, fill joints with mortar using trowel, and smooth.

Natural stone is a sympathetic material, full of character and available in as many shades and textures as there are types of quarried stone. New, it literally costs the earth.

Stone flags are the traditional flooring for country houses and cottages, especially in kitchen, hall and utility areas, but are rarer in new homes. Despite being impervious, tremendously hard-wearing and attractive, such floors are also heavy, cold, hard underfoot and need very strong sub-floors. Types include warm coloured sandstones, grey York stone, granite setts, creamy Portland stone, porous limestone and reconstituted stone slabs, as well as stone chippings mixed with cement and cast in slabs. These should be laid in mortar indoors, over a damp-proof membrane. If you inherit a flagstone floor, you may need to lift, damp-proof underneath, clean the slabs and replace. Sweeping and scrubbing are all the maintenance the floor should need, but porous types may stain and can sometimes be sealed – get expert advice.

Patios, porches, greenhouses, conservatories and paths are all suitable locations for stone floors. Thin sliced stone could be interesting for a roof garden (but check the load-bearing capacity of the roof first). Lay garden stone in sand: its own weight will keep it in place. Ensure foundations are firm and well-drained and that slabs are settled evenly on foundations. Fill joints with mortar, sand, gravel or earth and plants. Secondhand stone, broken or cracked, is a cheap alternative, suitable for crazy paving. This should not look too crazy – jigsaw it together neatly and with evenly sized joints.

Granite

Granite is a very hard rock – often dark blue-grey or mottled white – available in many colours, with a rich texture and a faintly crystalline appearance. Surfaces are often slightly irregular but can be honed, polished, flame-textured or sawn.

Granite is also available in paving stones known as setts. These can be laid in similar patterns to bricks although they are larger and rather more uneven, and also look good laid randomly or in irregular shapes. Granite is maintenance free but, if polished, can be washed and buffed.

Slate

Slate is a beautiful, evocative material which, bizarrely, can manage to look bogus if badly handled – especially if used in too small an area. It has a slightly rippled surface and comes in deep greys, grey-blues and grey-greens, though slightly reddish shades also exist. Quarried in mountain regions, it is very durable if properly cut: it can be sliced into thin layers, varying from about 1 to 10cm (½ to 4in). Although expensive, cold, brittle, heavy, hard underfoot and noisy, it makes a uniquely dramatic floor, a stunning foil for rugs.

Lay it in a concrete bed on a concrete sub-floor. Slate can be polished but it's best to leave it alone. Because of its slow water absorption slate is particularly good for outdoor use.

Left: A 'carpet' made of stone – large flagstones, smaller pebbles – works well in this country-house kitchen. The central ring of pebbles marks out and gives emphasis to the dining area.

Above: Two different shades of natural stone, laid in a pattern, focus the attention on the central working/eating area. The pleasant, overall effect is one of controlled informality.

Minimal marble used to achieve maximum effect. The main 'language' of the flooring is quarry tiles, with a black marble threshold mat separating one area of tiling from another, and inlaid marble strips providing physical and visual links.

Marble

Marble is a hard crystalline form of limestone which comes in a wide range of colours and patterns. It is expensive because of its weight and the difficulty of transporting and cutting it. Often used in hotels or office foyers, it has acquired a slightly commercial flavour. Domestically, it is grand and opulent but hard, cold and noisy underfoot. It looks good in dramatic patterns – black and white contrasted, for example – but it is very much a supplier's job to specify and lay.

The most expensive way to lay this material is in large sheets, which should be 16mm (5/8in) thick for internal use, but marble nowadays comes in many tile forms, often backed with other materials such as epoxy resin and fibreglass, or expanded polyurethane and bonded steel, which means the marble itself need only be quite thin. Cheaper, more practical versions are also available, which consist of marble chippings mixed with polyester resins and then ground and polished smooth. Marble must be installed on a floor that can bear its weight and normally one screeded with sand and cement.

While durability is high, marble is not completely stain-resistant. Mild abrasive detergents may be adequate for cleaning pale marbles but no alkaline soaps or sodas should be used on polished marble: they will remove the polish. Bad stains may need to be removed by specialist cleaning.

Terrazzo

Marble or granite chips mixed with concrete, cement or resins, plain or coloured, and ground smooth, terrazzo comes in tiles or slabs or is mixed and laid by trowel *in situ.* A practical, durable and water-resistant material, it makes a hard and noisy floor but is popular in commercial interiors. Because of its strength it can be laid as thinly as 9mm (3/8in).

The trowel-laid method of installation has spread to other mixtures of stone chips and various compounds and resins – one range consists of river-bed chips dye treated to produce an amazing range of colours and a surface like slightly craggy carpet. Terrazzo is expensive to install domestically and the job has to be done professionally. To maintain, sweep and wash, do not use strong acid or alkaline solutions.

Left: This 'outdoor carpet' is made of largish pebbles laid in circular patterns within a framework of brick latticing and border.

Below: When planting in gravel, brush gravel gently to one side until earth is exposed. Dig hole and plant. Press earth gently back into place. Rake over.

Cobbles and pebbles

You can collect your own pebbles, but choose stones that are of reasonable size, smooth and even. Cobbles, larger and more egg-shaped, were an early form of paving. They are fairly cheap and can be bought at garden centres.

Both pebbles and cobbles need to be bedded deep and flat (in clay, or in a mortar of one part cement to three parts sand) in order to make a surface comfortable enough to walk on. When laying in mortar, lay them wet so they do not absorb water from the mortar and so weaken it. Make sure they are close together so mortar is not visible; grout tactfully.

The garden is their natural home, but pebbles and cobbles could look very effective in the courtyard or lobby of a holiday cottage. In gardens they can be fixed for paths or edges, or left loose around trees or ponds. Bedded with larger stones, pieces of wood, and sculpture, they could make a 'hard' garden, or form the basis for rambling plants.

Gravel

Gravel comes in a huge variety of sizes, colours and types of stone. It makes a cheap surface, but should not be laid too deep or it will be difficult to walk on. It can look elegant, neat and natural especially when raked, Japanese-style; shades vary from dazzling white (useful in a dark garden) to dark colours, so it can be used to create pattern either by itself or in conjunction with other garden materials.

Lay it to a depth of 2.5cm (1in) into the same depth of a binding material. For a thicker layer, cover 10cm (4in) of hardcore with a thin layer of sand or fine aggregate, and lay gravel on the top. However you lay it, a fall of one in thirty and very even raking is necessary to avoid puddles.

The problem with gravel is that it gets everywhere – into the house, out into the street, on flowerbeds and lawns. Children also love to throw it around; cats may adopt it as a site for littering.

BRICK

Flooring bricks must be the type suitable for paving, otherwise the surface will quickly degrade. Either clay or calcium silicate bricks will do. While bricks are normally rectangular in shape, sizes vary and thicknesses range from the standard 6.5cm (2½in) to thin paviors of 2.5 to 5cm (1 to 2in). Colours include warm buffs, goldens, greys, reds, browns, and cool blues, greens and grey shades, as well as speckled mixtures. Surface texture can vary as can porosity – some bricks can absorb as much as 20 per cent of their own mass. Brick is a reasonably priced material and is available secondhand, but check that quality is appropriate and clean before use.

Indoors, brick makes a durable, but hard and noisy floor. It looks equally good right through the house, in hall, kitchen and sitting room, from the front door to the patio.

Lay it on concrete cured for a minimum of 28 days, on a damp-proof course on a ground floor, into a stiff mortar bed 2.5 to 3cm (1 to 1¼in) deep, with mortar jointing carried out simultaneously. Absorbent bricks should be briefly ducked in water before laying so that they don't suck too much water out of the mortar. Don't seal, oil or polish. Sweep, wash with mild detergent and rinse carefully.

Outdoors, brick is one of the most sympathetic materials for town and country garden alike. It can look crisp and regular or softer and more rustic, depending on how it is used. As well as paths, patios and terraces it is also suitable for slopes, steps, walls or mixed with other materials. Thin paviors can be used on roof terraces. Make sure outdoor bricks are frost-resistant.

Outdoors it does not need to be laid as precisely as indoors. Lay on a bed of firm even sand 2 to 4cm (¾ to 1½in) deep, on top of a layer of consolidated hardcore or ash. Margin bricks can be bedded in mortar to keep stretches of brick fixed firmly in place. Sand-laid bricks need to be pointed, a fiddly procedure which can stain the brick surface. Try a dry sand and cement mix, brush between joints, clean off surplus, water and allow to set. Don't be tempted to over-fill the joints: they really look best scooped lightly below the brick surface, so that the different shades of the individual bricks stand out cleanly.

Laying Bricks Outdoors

1. Set out dimensions with pegs and string, checking that the corners are right angles.

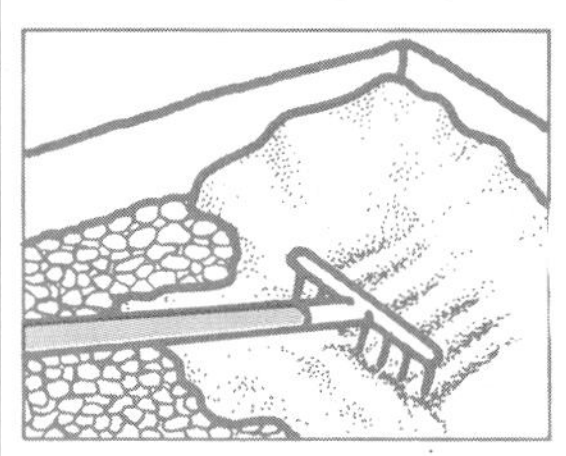

2. Dig to required depth. Backfill with consolidated hardcore. Sand. Rake level.

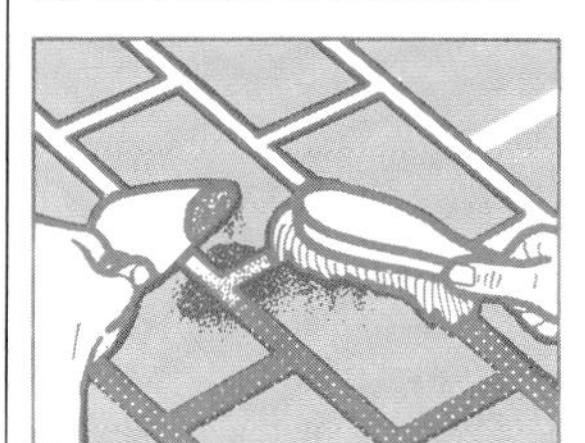

3. Lay bricks with narrow joints. Carefully brush in a dry mix of cement and sand.

4. Using a watering can with a fine rose, or a fine spray from a hose, water the brickwork to set the mortar.

With their warm colour and inherent variations, old bricks bring charm to any environment. This stretcher bond pattern unifies this whole area, and has a pleasant, soothing look.

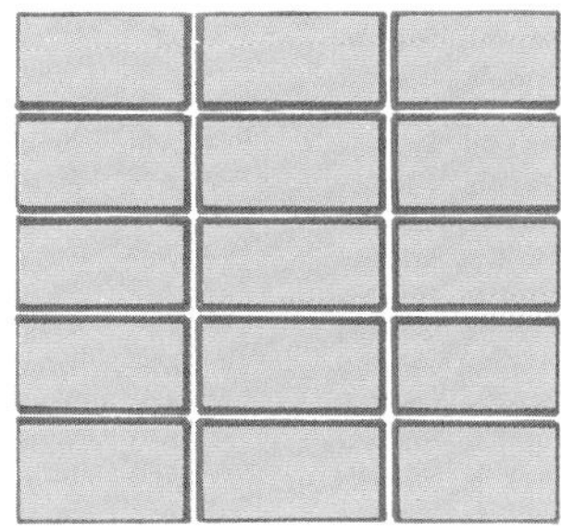

Stack bond *Bricks in straight rows. Easiest of all to lay. Ideal for small, unfussy areas.*

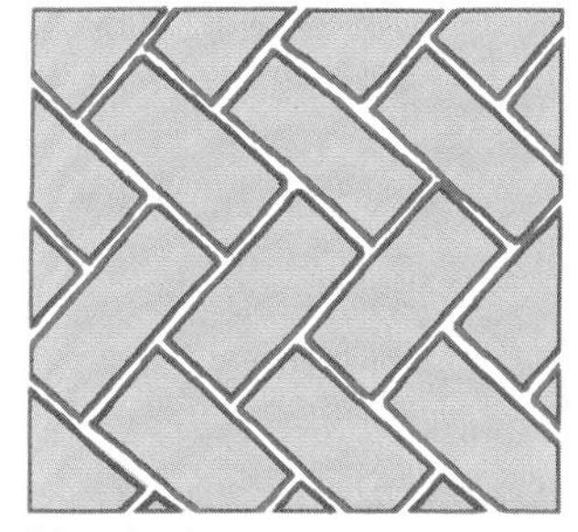

Herringbone *A very popular pattern. Pleasing to look at, it is also easy to lay.*

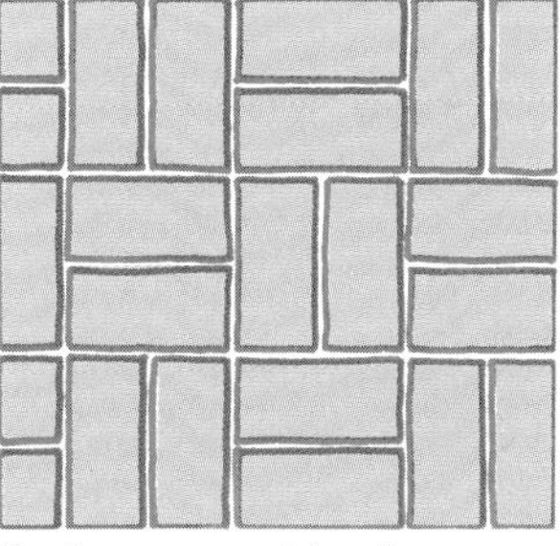

Basket weave *Visually pleasing. Somehow manages to look inwards. Good as a path.*

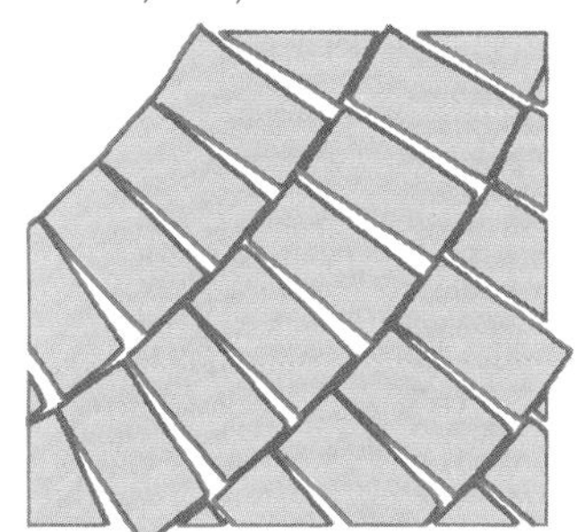

Radial *Ideal as a surround to curved or circular features. Comparatively easy to lay.*

Patterns

Part of the charm of brick derives from its human scale, which also enables a variety of patterns to be created easily. These include bond patterns, herringbone and basket-weave. Patterns are partly dependent on whether brick is laid on its bed (largest) face or on an edge. Fewer bricks will be needed for face laying, which gives a proportion of length to width of two to one as opposed to three to one of edge laying, and therefore demands forty standard bricks as opposed to sixty for each square metre (yard).

INDEX

Acknowledgments

The publishers would like to thank: Dianne Amerasekera; John Boddy Timber Ltd; Covent Garden General Store; Habitat; Heal's; Kibble, White and Blackmoor Ltd; Oriental Carpet Manufacturers (London) Ltd; Sadolin (UK) Ltd; Tile Mart Ltd; Tomkinsons Carpets Plc

Illustrators: Linda Broad; Gerrard Brown; Richard Phipps; Rob Shone; Elsa Wilson

Picture credits:
Abbreviations: CO – Conran Octopus Ltd; EWA – Elizabeth Whiting & Associates; *MMC – La Maison de Marie Claire; WOI – The World of Interiors*

Abitare/Gabriele Basilico 18-19, 30; Alnolog 11; *Ambiente*/ Sebastian Geifer Bastel 71; Guy Bouchet 73; Camera Press 6-7, 13 above, 42-3, 44-5, 68-9, 74; *House & Garden*, US François Halard 12; CO/Simon Brown 35, 54; CO/John Heseltine 22-3, 36, 37, 40, 41, 47, 48, 49, 50, 51; CO/Peter Mackertich 55, 66; Conran's 46; EWA/Clive Helm 31 right; EWA/Neil Lorimer 56; EWA/Michael Nicholson 17 right; EWA/Spike Powell 18; EWA/Tim Street-Porter 17 centre, 76-7; Futon Company 20; *Good Housekeeping*/Jan Baldwin 57, David Brittain 9; Susan Griggs Agency/Michael Boys 69; Habitat 15 right, 38, 41, 60, 62, 63 right, 67; Pat Hunt 40; Ken Kirkwood 13 below, 14, 21; *MMC* 10-11, 75; *MMC*/Jean-Luc Eriaud 19 right, 64-5, 70; *MMC*/Claude Pataut 16; *MMC*/ Antoine Rozes 51; Bill McLaughlin 33; Jeffrey McNamara 65; Sheppard Day 63 left; Syndication International 19 below; Agence TOP/Pascal Hinous 15 left; Transworld Features (*Casa Brava*) 61; *WOI*/Clive Frost 17 left, 31 left; *WOI*/Jean Bernard Naudin 72-3